Heterogeneous
Reaction
Dynamics

Heterogeneous Reaction Dynamics

Steven L. Bernasek

Steven L. Bernasek
Department of Chemistry
Princeton University
Princeton, NJ 08544-1009

This book is printed on acid-free paper.

Library of Congress Cataloging-in-Publication Data
Bernasek, S. L. (Steven L.)
    Heterogeneous reaction dynamics / by Steven L. Bernasek.
      p. cm.
    Includes bibliographical references and index.
    ISBN 0-89573-742-6 (alk. paper)
    1. Molecular dynamics.   2. Chemical reaction, Conditions and laws of.   3. Gas-solid
interfaces.   I. Title.
    QD461.B45   1995
    541.3'93—dc20   95-14954
                                                                      CIP

© 1995 VCH Publishers, Inc.

Printed in the United States of America

ISBN 0-89573-742-6 VCH Publishers, Inc.

Printing History:
10 9 8 7 6 5 4 3 2 1

Published jointly by

VCH Publishers, Inc.   VCH Verlagsgesellschaft mbH   VCH Publishers (UK) Ltd.
220 East 23rd Street   P.O. Box 10 11 61   8 Wellington Court
New York, New York 10010   69451 Weinheim, Germany   Cambridge CB1 1HZ
                                                                   United Kingdom

FOR SANDY, LISA, AND ERIC

Preface

In their book *Molecular Reaction Dynamics,* R. D. Levine and R. B. Bernstein describe their subject as being "concerned with the molecular level mechanism of elementary chemical reactions." This book, first published in 1974, served as a primer for students and researchers interested in the developing study of the detailed dynamics of gas phase chemical reactions. This field of research grew primarily because of the development of experimental techniques such as molecular beam scattering and laser spectroscopic methods. Recent experimental and conceptual advances have also made possible the development of the field of heterogeneous reaction dynamics, the subject of the present volume. Heterogeneous reaction dynamics, defined in analogy with the subject of Levine and Bernstein's primer, is the study of the molecular level mechanism of elementary chemical reactions occurring at interfaces between two phases. The scope of this book limits this definition to studies of elementary chemical reactions at the gas–solid interface, and also will hopefully serve as a primer for students and researchers interested in this growing area.

This book is directed to upper level undergraduates and beginning graduate students in chemistry and molecular physics, who would like an introduction to the rapidly growing literature of this field. As treated in this book, the field is concerned with the detailed dynamics of chemical reactions occurring on well-characterized solid surfaces. The book is organized around case studies from the literature, and the choice of these case studies is restricted to investigations that satisfy two criteria. The first is that the surface involved is well characterized as to structure and composition. This usually means single crystal substrates prepared and maintained in ultrahigh vacuum. The second criterion is the molecular level characterization of the gas phase participant in the heterogeneous reaction. The tools and approaches of molecular reaction dynamics, as defined by Levine and Bernstein, figure prominently in this criterion. This book is not meant to be an exhaustive research monograph, with complete coverage of every system that has been studied that meets these criteria. Rather, it is meant to illustrate the sorts of studies that can be done, and to illustrate the various aspects of heterogeneous reaction dynamics ranging from energy transfer in the gas-surface collision to rather more complex chemical reactions at surfaces.

Following a description of surface characterization methods and reaction dynamics approaches that are common to all of the case studies that make up this book, several chapters are presented that deal with heterogeneous reaction dynamics broadly defined. Each individual chapter begins with a discussion of experimental methods that are particular to the case studies described in that chapter. A number of case studies are discussed throughout the book that illustrate various aspects of heterogeneous reaction dynamics and that are exemplary of work going on in the field presently. These case studies begin with a discussion of inelastic scattering of molecules from surfaces, and the problem of energy transfer on collision. The processes of adsorption, film growth, and adsorbate interactions are then discussed. Following adsorption, the adsorbates involved in a surface reaction often must diffuse over the surface before reaction or subsequent desorption. This diffusion process has been studied in great molecular detail, and forms the basis for several interesting case studies. Dissociative adsorption of small molecules on initial collision with the surface is also an important possible step in a heterogeneous reaction. Studies of this process are described in Chapter 6. Atom recombination on surfaces is probably the simplest true chemical reaction that has been studied in complete molecular detail. Case studies of this process provide excellent examples of early and more recent work in the field of heterogeneous reaction dynamics. Catalytic oxidation, particularly the oxidation of CO, is a very well studied system that provides case studies illustrating the complexity of surface reactions that have been investigated. The final chapter of this book discusses small molecule decomposition processes. These reactions are at the edge of what it is possible to study using the tools of heterogeneous reaction dynamics, and provide the systems that will form the bulk of molecular level surface reaction dynamics studies in the future. Each chapter closes with a brief section of comments on where the particular subfield is going, and what specific questions remain unanswered in these studies.

The idea for this monograph grew out of discussions with Dr. Ed Immergut of VCH Publishers at an American Chemical Society National Meeting in Washington, D.C. in 1987. At that meeting I had organized a symposium dealing with the topic of heterogeneous reaction dynamics, and many of the investigators whose case studies make up this book were participants in that symposium. Over the years since those initial discussions, the case studies presented here formed parts of two graduate level special topics courses I gave at Princeton University. I thank the students in those courses for their comments and their interest in the material that makes up this book. Throughout my career at Princeton, my own research work in heterogeneous reaction dynamics, several aspects of which appear in the case studies described here, has been generously supported by the Chemistry Division of the National Science Foundation. I would also like to acknowledge the financial support of the Alexander von Humboldt Stiftung, and the hospitality of Professor George Comsa at the Forschunszentrum Jülich during a sabbatical leave in 1990, when an initial draft of this book was prepared. I also appreciate very much Ed Immergut's continued interest and patience during the preparation of this monograph. Finally, I wish to acknowledge the undergraduate, graduate, and postdoctoral students who have worked with me at Princeton over the years. Their contributions to the ideas and the data that make up a great deal of the work described here are essential.

Steven L. Bernasek
Princeton, New Jersey

June 1995

Contents

1

Introduction

Over the past four decades, a convergence of experimental techniques and theoretical methodology has resulted in the detailed understanding of a number of gas phase chemical reactions. The details of this understanding have included not only the knowledge of elementary mechanism and the ability to measure specific elementary rate processes, but also knowledge of quantum state-specific reaction probabilities and the charting of large regions of the interaction potential energy surfaces governing these reactions. Detailed comparisons have become possible between quantum state specific measurements and solutions of the Schrödinger equation for prototypical reactions such as the hydrogen exchange reaction $H + H_2 \rightarrow H_2 + H$. This field of study, described broadly as *molecular reaction dynamics,* has played a dominant role in basic chemical physics and physical chemistry research over this time. A number of excellent books and research monographs have addressed this field, and provide a good picture of the excitement of work in this area and of the development and scope of the field.[1]

One theme that is quite clear from these monographs is that the development of the understanding of gas phase molecular reaction dynamics was and is very strongly tied to developments in experimental and theoretical technique. Molecular beam scattering methods, first measuring total scattering cross section, then differential cross sections in universal detector crossed beam machines contributed enormously to this development. The invention of the laser and its application to problems in chemical dynamics, refinements and developments in spectroscopy, application of methods for molecular alignment, developments in classical and semiclassical scattering theory, and improvements in calculational

1

algorithms and hardware all contributed and interacted to provide the sort of detailed phenomenological description of gas phase reaction dynamics on a molecular level that is typical of frontier research in this area.

As a result of this detailed, state-specific description and theoretical interpretation of molecular reaction dynamics, the field has developed to the point where useful reaction dynamics rules can be stated. Ideas such as steric effects and energy disposal in chemical reactions can be placed on a quantitative footing and be directly explored. State-specific excitation of molecular species can be carried out, and bond-specific chemical reactions can be explored from a viewpoint of a rather clear understanding of the details of molecular motion.

Over the past two decades a somewhat similar scenario has been developing in the study of chemical reactions that occur on solid surfaces, rather than entirely in the gas phase. This developing field of *heterogeneous reaction dynamics* is the subject of this book. As with gas phase molecular reaction dynamics, this field owes its development largely to the emergence of experimental and theoretical methods that have made it possible to address questions of surface reaction mechanism and dynamics on a molecular level. The discussion in this book will concentrate on the study of heterogeneous reaction processes on a molecular level, and on systems that can be considered well characterized on that level.

Heterogeneous reaction dynamics, as discussed here, will be defined broadly to include the elementary dynamics of energy transfer and reaction at well-characterized solid surfaces. We will consider energy transfer between adsorbate and substrate and between gas phase and substrate, epitaxial growth and adsorbate interactions, diffusion on the surface, the adsorption process itself, as well as simple surface chemical reactions. Figure 1.1, an updated version of an often-used schematic, illustrates the range of processes that can and do occur when a gas interacts with a solid surface.[2] An attempt to understand this rich complexity on a molecular level is the subject of heterogeneous reaction dynamics.

It is evident immediately that the study of heterogeneous reaction dynamics is a much more complex undertaking than the detailed study of gas phase molecular reaction dynamics. This is because of the enormous complication, both experimentally and theoretically, that the solid surface introduces. Now, instead of dealing with the detailed motion of two or a few or several atoms as they make and break bonds or transfer energy in gas phase collisions, the motion of these several atoms must be considered in combination with the behavior of the 10^{23} atoms that make up the solid, or at least with the region of the solid surface that can be expected to participate in the dynamic process. In addition to the two phase heterogeneity of the gas–solid interaction, the physical heterogeneity of the solid surface itself must be dealt with, where even ''ideal'' surfaces are far from perfect and small quantities of defects and imperfections may be controlling what is observed.

Because of this complexity, advances in understanding in the field of heterogeneous reaction dynamics are very closely tied to advances and developments

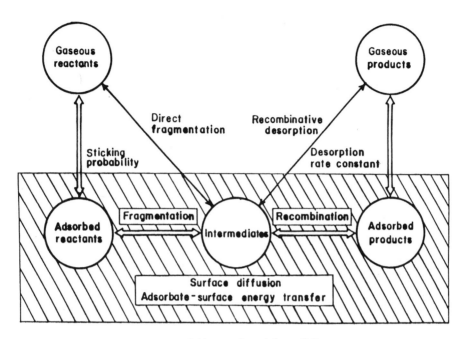

Figure 1.1. Schematic of gas–solid interaction. After ref. 2.

in experimental technique. Methods for the characterization and investigation and control of the solid surface must be combined with methods for the characterization and investigation and control of the gas phase reactants and products. This combination and interplay make this a very fertile area for technique development and application. It is important to recognize that understanding in this field relies on detailed knowledge of experimental approaches and clever application of these methods to probe the behavior of *both* the gas phase and surface participants in the heterogeneous dynamics process.

The following pages will provide a sampling of this exciting and fast growing field of research. The discussion is based on a number of exemplary "case histories" taken from the literature, including, but not limited to work in which the author has taken part. The examples discussed are all from areas actively being researched, and as such are bound to be incomplete as to description and inclusion of the very most recent relevant information. The intent is to illustrate important questions in the study of heterogeneous reaction dynamics, not to present an exhaustive literature survey. The examples presented are also likely to display a bias toward experimental measurement and phenomenological description, first because the author is an experimentalist, and second because theoretical understanding in this field is not so well developed. Heterogeneous reaction dynamics has not progressed yet to the point where detailed state-to-state measurements can be compared routinely with detailed theoretical predictions even for the simplest systems. Nor are reaction rules for heterogeneous reactions, gas-surface interaction potentials, or generalized concepts regarding surface reaction dynamics widely recognized or agreed on. One hopes that a monograph such as this prepared in 20 years would be able to address these points. For now, experimental and theoretical methods can be presented and heterogeneous reaction dynamics examples can be described. From these examples and further studies stimulated by their consideration, an understanding deeper than phenomenological should develop.

The remainder of this book is organized into eight chapters of various lengths. The first of these provides some basic background information concerning the preparation, characterization, and control of the solid surfaces involved in heterogeneous reaction dynamics studies on well-characterized solid surfaces. The remaining chapters are organized around particular subtopics in the general area of heterogeneous reaction dynamics. In each chapter, experimental methods specific to the study of that subtopic will be briefly described. Then illustrative examples of studies taken from the literature will be presented to indicate the phenomenology of this subtopic of heterogeneous reaction dynamics and to indicate the present level of understanding of the particular area. The presentation will necessarily be rather brief, with extensive references to the literature provided to guide the reader to the details of experimental procedure and interpretation. Finally, each chapter will provide some of the author's opinions on what is missing and where work might be fruitfully carried out in the future. The specific subtopics to be addressed in the following chapters are gas-surface and adsorbate-surface energy transfer, epitaxy and adsorbate interactions, surface diffusion processes, dissociative adsorption, atom recombination on surfaces, catalytic oxidation reactions, and small molecule decomposition processes. Each chapter will conclude with a summary of the author's view of the state of understanding of heterogeneous reaction dynamics, and reiterate some of the areas requiring future effort.

References

1. R. B. Bernstein, *Chemical Dynamics via Molecular Beam and Laser Techniques,* Clarendon Press, Oxford, 1982.

   P. R. Brooks and E. F. Hayes, eds., *State-to-State Chemistry,* ACS Symposium Series 56, American Chemical Society, Washington, D.C., 1977.

   W. H. Flygare, *Molecular Structure and Dynamics,* Prentice-Hall, Englewood Cliffs, NJ, 1978.

   H. S. Johnston, *Gas Phase Reaction Rate Theory,* Ronald Press, New York, 1966.

   R. D. Levine and R. B. Bernstein, *Molecular Reaction Dynamics,* Clarendon Press, Oxford, 1974.

   R. D. Levine and R. B. Bernstein, *Molecular Reaction Dynamics and Chemical Reactivity,* Oxford University Press, Oxford, 1988.

R. D. Levine and J. Jortner, eds., *Molecular Energy Transfer,* John Wiley, New York, 1976.

G. Scoles, ed., *Atomic and Molecular Beam Methods,* Vols. 1 and 2, Oxford University Press, Oxford, 1988, 1992.

I.W.M. Smith, *Kinetics and Dynamics of Elementary Gas Reactions,* Butterworths, London, 1980.

2. M. J. Cardillo, *Annu. Rev. Phys. Chem.* **32,** 331 (1981).

CHAPTER

2

Surface Characterization Methods

To hope to understand a heterogeneous reaction in molecular level detail, it is essential to be able to define the solid surface with atomic detail, and to maintain the surface as it has been characterized throughout the course of a measurement. This requirement implies the use of single crystal solid samples, cut to expose a particular crystallographic plane of interest, prepared and characterized in situ, and maintained in ultrahigh vacuum or other carefully controlled atmosphere during a measurement. The need for ultrahigh vacuum is readily seen by considering the fact that at a pressure of 1×10^{-6} torr, the flux of molecules to a surface exposed to this pressure is on the order of 1×10^{15} molecules/cm^2 sec. For a typical solid, surface atom densities are on the order of 1×10^{14} to 1×10^{15}/cm^2, indicating that every atom in the surface will be struck by at least one molecule per second from a gas phase pressure of 1×10^{-6} torr. If a reasonable fraction of those molecules incident on the surface adsorb, then a pure sample free of adsorbed molecules will stay that way for only a few seconds at a pressure of 1×10^{-6} torr. Thus, the effort to provide experimental systems with sample chamber base pressures in the 10^{-10} torr range and below. In fact, the modern study of surface physics and chemistry, and the field of heterogeneous reaction dynamics, is strongly linked to the development of techniques for attaining and maintaining ultrahigh vacuum conditions ($<10^{-9}$ torr). These techniques have been well described in a number of useful textbooks and handbooks,[1] and will not be described further here. The generation of a clean very low-pressure working environment is, however, essential to the studies that will be described in this book.

Once a clean working environment is established, facilities must be available for preparing and characterizing the sample in situ. In the work described in this book, the solid surface under study is generally a macroscopic slice of a high-purity single crystal of a metal, semiconductor, or insulator. The actual procedure for orientation, cutting, mechanical polishing, and chemical cleaning of the sample before mounting in the experimental chamber is very sample dependent. Descriptions of these procedures are contained in the literature[2] and are readily developed and adapted for new sample surfaces. The sample so prepared is then mounted in the UHV environment, generally on a manipulator that allows positioning and temperature control of the surface.

The mounted sample surface must now be further cleaned and prepared in situ. Several general methods for cleaning surfaces for heterogeneous reaction dynamics studies have been developed. The actual choice of method is again very sample dependent. In the case of refractory metal samples, very high temperature annealing in vacuum may be useful for preparing a clean surface. Often chemical treatments, such as annealing in oxygen or hydrogen, are effective in removing surface contaminants. Care must be exercised in this case to prevent direct deep oxidation or reduction of the sample itself, forming an oxide or hydride or irreversibly altering the composition of a compound sample. Another widely used technique is ion bombardment of the sample surface. In this case, energetic inert gas ions (usually Ar) are accelerated at the surface. The collisions sputter material from the surface, removing adsorbed layers and impurities segregated at the surface. Choice of ion, ion energy, angle of incidence of the ion beam, and ion current to the sample surface all vary the effectiveness of this cleaning procedure. Sputtering is then followed by annealing of the sample in vacuum to remove the residual damage to the surface caused by the sputtering process, and to desorb implanted inert gas atoms. Sample cleavage or fracture in the UHV system can also be used to prepare a clean surface for study. This is especially useful for semiconductor and insulator samples whose surface stoichiometry may be irreversibly changed by the methods described previously. It should also be kept in mind that these methods remove impurities only from the surface region of the solid. Annealing in vacuum or high temperature reduction and oxidation often result in the segregation of bulk impurities such as carbon or sulfur at the surface. These must then be removed by cleaning, annealing cycles until the bulk is purged of the impurity. This can often be quite a long and tedious process.

The results of in situ preparation of the sample surface must be monitored by in situ techniques that provide information about the geometric structure, composition, and electronic properties of the surface. An enormous number of methods have been developed to provide this sort of information. There are again a number of excellent textbooks and research monographs[3] that describe in detail the alphabet soup of acronymed methods available to the surface scientist. Brief descriptions of some of the most widely used methods and references to the literature for further information about them are provided in the remainder of this chapter. The methods discussed here will be classified somewhat arbitrarily as geometric structure, surface composition, and surface electronic structure

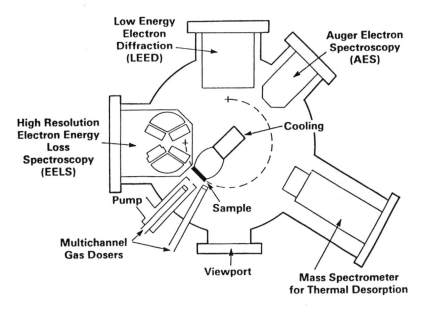

Figure 2.1. Schematic diagram of typical multiprobe ultrahigh vacuum apparatus for surface studies.

methods. A brief experimental description will be provided, along with an idea of the type of information that is available from the method. It must also be noted that these methods are often (in fact must) be used in combination. A single method rarely gives enough information for an adequate characterization of the solid surface. This multitechnique approach, which will be seen throughout the examples discussed in this book, is illustrated by the schematic diagram of a typical multiprobe UHV apparatus shown in Figure 2.1. This system includes low-energy electron diffraction (LEED) for geometric structure characterization, Auger electron spectroscopy (AES) for compositional characterization, high-resolution electron energy loss spectroscopy (HREELS) for vibrational characterization of adsorbed intermediates, and a quadrupole mass spectrometer for analysis of the residual gas in the UHV system and for detection of species desorbed from the sample surface during thermal desorption spectroscopy (TDS) measurements. In addition, an ion source for cleaning the sample and dosers for exposing the sample to controlled amounts of adsorbates are evident in the diagram. The techniques mentioned above, along with a number of other widely used methods, are described in what follows.

2.1 Structural Information

A knowledge of the geometric structure of the surface under study is essential to the understanding of heterogeneous reaction dynamics. For single crystal samples, as described above, the long-range surface periodicity allows the use of diffraction methods to provide geometric structural information. Such diffraction methods include LEED, reflection high energy electron diffraction (RHEED), small angle X-ray scattering, and atomic beam diffraction. All of these methods use diffraction of incident radiation (electrons, X-ray photons, helium atoms) to provide information about the atomic geometry of the surface region. Helium atom diffraction is the most surface sensitive of these probes, as thermal energy helium atoms do not penetrate the surface of the solid, and so provide information only about the periodicity of the outermost layer. LEED, RHEED, and X-ray scattering all have finite penetration lengths into the solids, so that the diffraction information is characteristic of the first several layers of the solid.

LEED is perhaps the most widely used surface structural probe,[4] due to its experimental simplicity and the relatively straightforward information it provides, at least on an initial level. For that reason, it will be described in some experimental detail in what follows. The other techniques mentioned are less widely used, but can provide much better information than LEED in certain instances. Helium atom scattering and RHEED, in particular, will be considered in more detail later in the context of specific dynamic examples. In addition, other techniques that provide structural information specific to the heterogeneous reaction dynamics examples to be considered below will be described along with the examples.

LEED is generally carried out in a display type apparatus, as indicated schematically in Figure 2.2. A low-energy, reasonably monoenergetic ($\Delta E/E \sim 0.05$) electron beam is formed by emission from a hot filament, collimated, and focused on the single crystal sample. Backscattered electrons are energetically

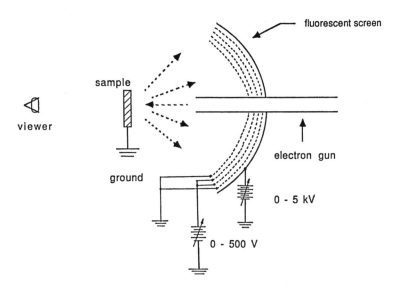

Figure 2.2.   Schematic diagram of low-energy electron diffraction apparatus.

filtered by a set of four hemispherical grids, and postaccelerated onto a phosphor-coated screen. This screen is then normally viewed through the window opposite. Elastically backscattered electrons are allowed to pass the set of grids, with those electrons inelastically scattered by the solid rejected by the grid optics. If the surface region exhibits periodic order, of dimension larger than the coherence width of the incident electron beam, then a pattern of diffraction spots is seen on the phosphor screen. The position of these diffraction features as a function of electron wavelength (voltage) provides information about the two-dimensional periodicity of the sample surface region.

On this level, the structural information provided by LEED is very straightforward. The spot pattern symmetry and the dimension of the pattern for a known voltage and apparatus geometry provide the symmetry and dimension of the two-dimensional real space surface unit cell. Changes in the spot pattern on adsorption or reaction on the surface provide rather direct information about changes in the symmetry and dimension of the surface unit cell. This level of information is readily obtained, and the experimental apparatus needed is modest.

More detailed structural information can be had with considerably more effort by an analysis of the intensities of the diffraction features as a function of the incident electron wavelength. The interaction of slow electrons with solids is a strong interaction, and electrons are likely to be elastically scattered more than a single time in the backscattering process. This multiple scattering, and the strong attenuation of the incident electron beam by the solid, dictate the use of a multiple scattering (dynamical LEED) formulation for the calculation of diffracted LEED intensities for comparison with measured intensities.[5] These calculations are quite computer intensive, and the use of LEED as a detailed surface structural probe requires the close interaction of experiment and theoretical calculation. In addition, the accurate measurement of diffraction intensities requires improvements to the display type LEED apparatus described above. Photographic recording and analysis of the diffraction intensities,[6] direct video readout of the intensities,[7] and direct measurement of the diffracted electron current have all been used to collect accurate LEED intensity data.[8] Surface crystallography using LEED has provided a large fraction of the detailed information that is available concerning the structure of surfaces and adsorbed layers.[9]

A recently developed surface structural tool, which does not rely on surface periodicity and diffraction conditions, is the scanning tunneling microscope (STM).[10] This technique provides a real space topographic display of the surface structure, often with atomic resolution. Although the method is so far not as widely used as LEED, new instruments are being developed and applied to problems of heterogeneous reaction dynamics at a rapid pace. The basis of the method is illustrated in Figure 2.3. A sharp probe tip mounted on a piezoelectric ceramic is rastered over the sample surface at a very small probe to sample distance. When a modest voltage is applied between sample and tip, tunneling will occur between them when the probe approaches the surface closely. The tunneling current is exponentially dependent on the probe–surface spacing and

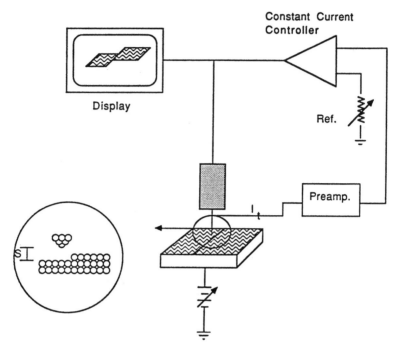

Figure 2.3. Schematic diagram of the basic scanning tunneling microscopy method.

can be used in a feedback control circuit to control the probe–sample distance by controlling the voltage applied to the piezoelectric holding the probe. If the microscope is operated so that constant tunneling current is maintained during the raster scan, then the voltage applied to the probe piezoelectric to maintain that constant current provides a topographic display of the surface structure. Such a display is shown for two surfaces in Figure 2.4, the Si(111) 7 × 7 surface, an ordered surface long studied by LEED and other surface crystallographic methods, and a platinum surface showing steps and growth island structures on a somewhat larger scale. Clearly, structural information is available from the STM method for samples of a widely varying degree of order and overall structure.

As with the brief discussion of LEED provided above, all of the advantages and disadvantages of STM as a structural probe cannot be described here. Certainly, consideration must be given to probe–surface interactions, local heating of the sample by the tunneling current, and the question of what electron density surface the tunneling profile actually probes. Also, the dependence of the topographic display on the size and polarity of the sample–probe bias voltage must be considered. STM is a very promising technique, which is still very much under development. It is likely in the future to provide answers to a number of crucial structural questions relevant to heterogeneous reaction dynamics.

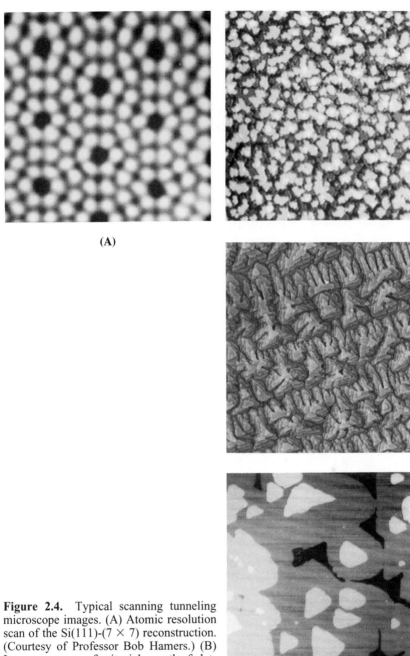

(A)

Figure 2.4. Typical scanning tunneling microscope images. (A) Atomic resolution scan of the Si(111)-(7 × 7) reconstruction. (Courtesy of Professor Bob Hamers.) (B) Large area scans of epitaxial growth of platinum on Pt(111) surfaces under various conditions. (Courtesy of Professor George Comsa.)

(B)

There are, in addition, many other surface-sensitive techniques that provide structural information. For example, vibrational spectroscopic methods such as reflection IR spectroscopy[11] or high resolution electron energy loss spectroscopy[12] (HREELS) can provide detailed information about structure and bonding in adsorbed molecular layers. Also, a number of synchrotron-based spectroscopic methods provide accurate detailed structural information for specific cases. These methods are pertinent to several of the heterogeneous reaction dynamics examples, and will be described in that context.

2.2 Compositional Information

Another class of basic information that must obviously be available for the detailed understanding of heterogeneous reaction dynamics is the actual atomic composition of the surface. This information can be obtained from a wide range of spectroscopic techniques, involving electrons, ions, and photons. Again, different methods are more or less suited for different surfaces or for different types of compositional information. Several methods will be mentioned in this section, with two of the most widely used techniques described in somewhat more detail.

The most widely used surface compositional probe is probably Auger electron spectroscopy (AES).[13] In this method, a beam of energetic electrons (2–10 keV) is used to excite secondary electron transitions known as Auger transitions in the solid. These transitions result in the ejection of electrons from the solid with kinetic energy characteristic of the electron energy levels of atoms in the solid, as illustrated in Figure 2.5. Thus, measurement of the secondary electron emission as a function of kinetic energy should provide information on the identity and relative amounts of various atoms in the surface region.

This simple statement is complicated somewhat by the fact that a number of other processes produce secondary electron emission, so that the Auger electrons are included on a much stronger background signal, as illustrated in the plot of Figure 2.6. For this reason, Auger spectra are most often recorded in a differential mode $[dN(E)/dE]$ to accentuate the small Auger peaks on the large secondary electron background. Also, the quantitative nature of the technique is not so obvious as the simple discussion above suggests. The number of Auger electrons of a particular energy emitted is certainly proportional to the number of a particular type of atom in the sampled region. The proportionality, however, depends on the energy of the transition, the matrix in which the atom is held, and a number of other factors that make absolute quantification difficult. Careful calibrations and standardizations must be used to obtain quantitative information.

In spite of these considerations for detailed quantitative analysis of the surface region, AES is extremely useful in determining the effectiveness of sample cleaning procedures. Especially for single component metal and semiconductor samples, AES can readily detect the presence of a few percent of a monolayer

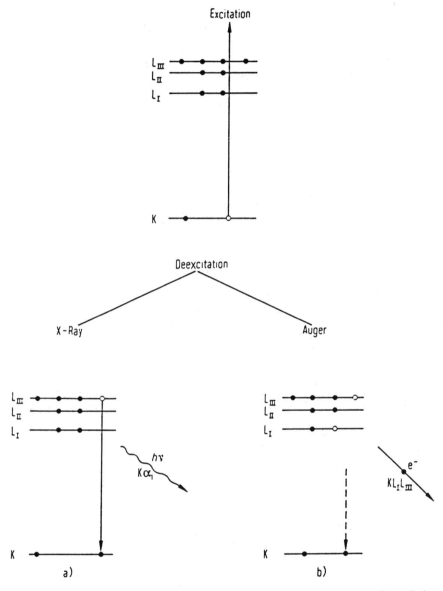

Figure 2.5. Energy level diagram for X-ray fluorescence (a) and Auger (b) emission following core hole state excitation.

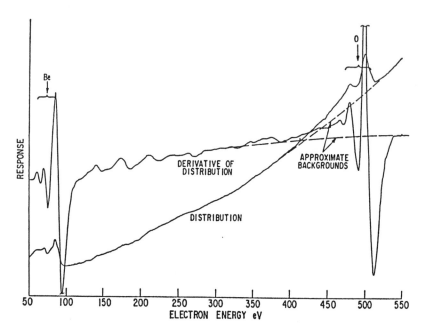

Figure 2.6. Example of Auger electron spectrum, showing the secondary electron distribution and the derivative of the distribution.

of common impurities, and can signify whether the sample is clean. For multicomponent systems, this becomes harder, but can still be accomplished with careful standardization and care in quantitative interpretation.

Experimentally, AES is relatively easy to implement, which also accounts for its widespread use. Auger transitions are normally excited by an electron beam of a few kiloelectron volts of energy. No special care need be taken to make the beam especially monoenergetic, as the Auger electron kinetic energy is independent of the exciting radiation. The emitted electrons can be energy analyzed in a number of possible ways. The more or less standard energy analyzer presently is the cylindrical mirror analyzer (CMA), which consists of concentric cylinders that electrostatically filter the emitted electrons.[14] This analyzer has quite a good electron throughput, reasonable resolution, and can be operated at high scan rates, making it a convenient and fairly sensitive device. The retarding field analyzer (RFA),[15] based on the four-grid LEED electron optics discussed earlier, is also widely used. It has the advantage that the same optics can be used for LEED and AES, reducing the cost and complexity of an apparatus used for heterogeneous reaction dynamics studies. It has disadvantages of resolution, sensitivity, and scan speed as compared with the CMA, however. Other electron energy analyzers are used for AES as well, such as higher resolution hemispherical analyzers, or compact 127° sector analyzers.

   The ability to finely focus and raster high-energy electron beams makes scanning Auger microscopy (SAM) possible as well.[16] In this case, a spatially resolved Auger image of the surface is obtained by scanning the excitation beam over the surface and imaging the electrons corresponding to specific Auger energies. Thus, an elemental mapping of the surface, with lateral resolution on the order of 2000 Å, is available. This sort of information is quite useful in device preparation, or technical surface analysis. For large, well-ordered uniform samples as are normally used in heterogeneous reaction dynamics studies, spatially dependent information on this scale is perhaps less essential.

   A second widely used technique for surface compositional determination is X-ray photoelectron spectroscopy (XPS).[17] In this technique, as with photoelectron spectroscopy in general, an energetic photon incident on a solid can eject a photoelectron, whose energy, given by (2.1)

$$E_{KE} = h\nu - E_{BE} - e\phi \tag{2.1}$$

is characteristic of the binding energy of the electron in the solid, for a given photon energy $(h\nu)$ and sample work function (ϕ). For photons of X-ray energy, the emitted electrons originate from core levels of the atoms in the solid. Again, a measurement of the number of photoelectrons emitted at a particular kinetic energy will provide information about the type and number of atoms in the irradiated region of the solid.

   As was the case with AES, the absolute quantitative analysis of the surface region is not completely simple. Variations in penetration depth of the exciting radiation for different materials, changes in excitation cross section depending on the chemical environment of the atom, and differences in electron mean free path in the solid as a function of energy all contribute to complicating the attempt at quantitative analysis. By the use of standard conditions and standard samples, these problems can be mostly worked out, and XPS proves to be a rather satisfactory quantitative tool for surface analysis.

   In comparison with AES, XPS methods are somewhat less convenient, since data collection times are generally much longer for XPS measurements. Also, XPS is somewhat less sensitive to low Z contaminants such as carbon, nitrogen, and oxygen than is AES, so its use for monitoring impurity levels during cleaning is less satisfactory. XPS also, in common with AES, samples the first several layers of a solid. It is not specifically first layer surface sensitive, as are techniques that use ions or low-energy atoms as probes.

   Experimentally, XPS has much in common with AES. The ejected photoelectrons must be energy analyzed and detected with reasonable resolution and sensitivity. The CMA or hemispherical energy analyzer is readily able to do this. The increased resolution, which is useful for analyzing peak positions in XPS, and thereby obtaining electronic structure (chemical) information about the surface, is normally obtained by the use of large hemispherical analyzers. The source for XPS, rather than a focused electron beam, is a standard X-ray source. Generally an Mg or Al anode is used, and the source is attached directly to the

UHV system. Since excitation cross sections are lower for XPS, direct pulse counting of the analyzed electrons is normally used, rather than lock-in detection of a substantial current as is the case in AES.

There are, of course, other approaches to the compositional analysis of surfaces. Energy analysis of elastically scattered medium energy ions, ion scattering spectrometry (ISS),[18] provides compositional information about the outermost layer of the solid. In this case, kinematic scattering of the detected ions is assumed, and their energy loss will then be inversely proportional to the mass of the surface atom suffering the collision. Counting the number of scattered ions at a particular energy then gives a quantitative analysis of the surface. Another ion probe of surface composition is secondary ion mass spectrometry (SIMS),[19] which mass analyzes the surface atoms sputtered away by an energetic incident ion beam. This is obviously a direct detection of the mass of atomic or molecular species present on the surface. SIMS has the disadvantage that the incident ion beam removes the top layer of the surface during analysis. If very low ion currents are used this problem can be minimized, however. Another serious problem, especially for analysis of molecular overlayers, is that the sputtering process is not understood well enough to conclusively link the presence of a particular fragment ion in the SIMS spectrum with its actual presence on the surface. Carefully used on atomic systems, though, SIMS has excellent sensitivity and can detect impurity levels well below those accessible to AES or XPS.

A number of other desorption techniques have been developed that can be used in favorable instances to provide surface compositional information. Electron-stimulated desorption (ESD),[20] photon-stimulated desorption (PSD),[21] laser-initiated thermal desorption (LITD),[22] and normal thermal desorption spectroscopy (TDS)[23] all provide signals that are indicative of the composition of the surface being probed. These methods are generally sensitive to molecular overlayers, or to minority species on the surface, but do not provide a general or universal compositional analysis.

2.3 Electronic Structure Information

Information about the electronic structure of surfaces can be obtained from a broad range of methods as well. The electronic structure of a surface is certainly connected strongly to its geometric structure and its atomic composition. The distinction made here concerns the chemical environment of the atoms on the surface. It concerns the electronic density of states of the perfect surface, and also modification of that electronic structure by defects or impurities or reactive sites on the surface. A detailed knowledge of this aspect of the surface is difficult to obtain, and adds to the complexity of describing the ''state-to-state'' dynamics of heterogeneous reactions.

Electronic structure information is certainly available from photoelectron spectroscopy. Ultraviolet photoelectron spectroscopy (UPS),[24] in particular, provides information about the valence band density of states (DOS) of the surface region. As with XPS, UPS measures the kinetic energy of photoemitted electrons. In this case, the excitation source is in the ultraviolet energy range, and electrons are ejected from within several electron volts of the Fermi energy. The photon source in this case is often an inert gas discharge lamp, usually a helium discharge lamp that has strong lines at 21 and 40 eV. Variable energy UV sources, such as synchrotron sources, can also be used to excite UV photoelectron spectra. Such high-intensity, monochromatized sources can provide a great deal of detailed information about the electronic structure of solids. They also have sufficient intensity to make dynamic measurements of electronic structural changes during the course of a heterogeneous reaction, although this application of synchrotron radiation has not yet been widely pursued. Figure 2.7 illustrates UPS spectra that show the effect of a surface reaction on the valence band density of states of a TiO_2 single crystal surface.

XPS spectra also contain useful information about the electronic structure of surfaces, as indicated above. The exact binding energy of electrons photoemitted from core levels of atoms in the solid is strongly affected by the chemical environment of the atom. To a first approximation the binding energy reflects the oxidation state of the atom emitting the photoelectron. Of course, the connection is not so direct, both because the oxidation state is a poorly defined quantity, and because the binding energy is affected not only by the initial state of the ionized atom, but also by the final ion state and how it relaxes energetically with respect to its neighbors in the solid. The binding energy information is still useful, in any case, and can be used to monitor reaction processes in heterogeneous systems, as is illustrated in Figure 2.7.

Another widely used probe of surface electronic structure is the measurement of the work function of the surface.[25] This is a surface averaged quantity, which corresponds to the energy required to take an electron from the Fermi level of the solid, and remove it to infinity. The value of the work function is changed on adsorption of an electron-donating or electron-withdrawing overlayer and such changes in electron density in the surface region can be monitored by measuring changes in the work function.

There are a number of experimental methods that can be used to measure work function change, $\Delta\phi$, on adsorption or reaction at a surface. One will be described here, and a number of other approaches mentioned. The most widely used method is the Kelvin probe.[26] In this method, a vibrating flat disk probe is brought close to the surface under study. The vibrating probe and the surface comprise a flat plate capacitor, whose capacitance is modulated by the changing distance between the plates. For fixed surface conditions, the charge stored by the capacitor is monitored by the ac voltage across the vibrating capacitor. When a molecular layer adsorbs on the sample surface, the capacitance of the system changes due to the change in work function of the sample surface. This is evident by a change in the measured voltage across the capacitor. When using this approach care must be exercised in the choice of probe materials. The probe must be constructed of a material whose work function does not change when

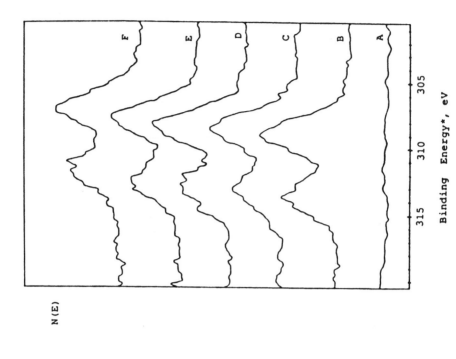

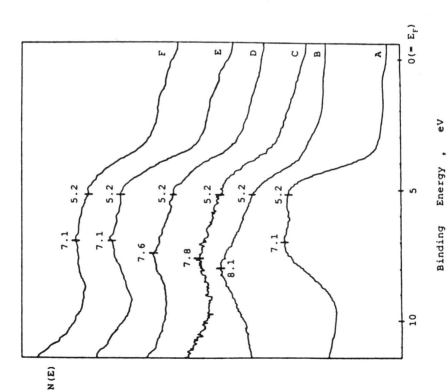

Figure 2.7. Ultraviolet and x-ray photoelectron spectra of a rhodium organometallic complex deposited on TiO_2, for various reaction treatments (A–F).

exposed to the various adsorbates under study. Often a gold surface or a well-oxidized tungsten surface is used for these probes.

Other methods for measuring work function change include the diode method,[27] where the change in bias voltage needed to repel a low-energy electron beam incident on the surface following adsorption is a measure of the work function change. This method is especially convenient because the standard four-grid LEED optics and LEED gun can be used to carry out the measurement. Photoemission can also be used to measure work function, as can a measure of thermionic emission current for a range of temperatures. Each of these approaches is appropriate to particular circumstances, such as the need for a measure of absolute work function in the first instance, or for refractory metal surfaces in the second.

In any case, the change in work function on adsorption provides information about the electronic structure of the surface. It indicates relative dipole layer orientation and is also useful as a measure of adsorbate coverage. Work function change is an averaged quantity, however, and does not provide the detailed electronic structural information available from photoelectron spectroscopy or other more specific techniques.

A very promising specific technique that has not yet developed to the level of routine use is the technique of scanning tunneling spectroscopy (STS).[28] This method, based on the STM technique described earlier, probes the electronic structure near the Fermi level for particular spatially resolved regions on the sample surface. This is accomplished by monitoring the tunneling current for fixed sample-probe distance, while varying the bias voltage. These I–V curves provide a measure of the occupied DOS below the Fermi level (sample negative, tip positive) and the unoccupied DOS above the Fermi level (sample positive, tip negative). Such I–V curves can in principle be collected at every point parallel to the surface in the scan region corresponding to an STM topographical scan. This can provide information about changes in local electronic structure near an adsorbed molecule, or at a defect site such as a step or missing atom.[29] An example of this sort of detailed electronic structural information is given in Figure 2.8, which illustrates a series of I–V curves for STS scans of an Si(111) surface with an adsorbed NH_3 overlayer.[30] It can be seen that electronic structural changes are evident in regions of the unit cell where NH_3 is bound to the missing atom sites on Si(111). These changes are readily interpreted as electron transfer indicative of molecular binding on the surface.

Only a few of the many techniques available for monitoring the electronic structure of solid surfaces have been mentioned here. Research monographs dealing with surface spectroscopy provide much more detailed information on the techniques already mentioned, as well as descriptions and evaluations of many more. The same is certainly true for methods of surface structural and surface compositional determination. More details about these methods are available in the literature, along with many other techniques. The general references provided in this chapter are a good place to start to learn about these methods. New approaches to surface characterization are being steadily developed, help-

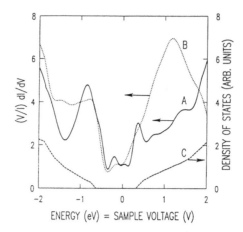

Figure 2.8. Scanning tunneling spectroscopy measurements on the Si(001) surface exposed to NH_3 (ref. 30).

ing to provide the detailed characterization of the solid surface that is essential to the understanding of heterogeneous reaction dynamics.

References

1. S. Dushman and J. M. Lafferty, *Scientific Foundations of Vacuum Technique,* 2nd ed., John Wiley, New York, 1962.

    N.T.M. Dennis and T. A. Heppel, *Vacuum System Design,* Chapman and Hall, London, 1968.

    A. Roth, *Vacuum Technology,* North-Holland, Amsterdam, 1976.

    P. A. Redhead, J. P. Hobson, and E. V. Kornelsen, *The Physical Basis of Ultrahigh Vacuum,* Chapman and Hall, London, 1968.

    "Vacuum Technology: Its Foundations, Formulae and Tables," in *Product and Vacuum Technology Reference Book,* Leybold-Heraeus, San Jose, California, 1986.

2. R. G. Musket, W. McLean, C. A. Colmenares, D. M. Makowicki, and W. J. Siekhaus, *Appl. Surf. Sci.* **10**, 143 (1982).

3. G. A. Somorjai, *Chemistry in Two Dimensions: Surfaces,* Cornell University Press, Ithaca, NY, 1982.

    G. Ertl and J. Kuppers, *Low Energy Electrons and Surface Chemistry,* VCH, Weinheim, 1985.

    D. P. Woodruff and T. A. Delchar, *Modern Techniques of Surface Science,* Cambridge University Press, Cambridge, 1986.

    R. B. Hall and A. B. Ellis, eds., *Chemistry and Structure of Interfaces: New Laser and Optical Techniques,* VCH, Weinheim, 1986.

    D. Briggs and M. P. Seah, eds., *Practical Surface Analysis,* John Wiley, Chichester, 1983.

    L. C. Feldman and J. W. Mayer, *Fundamentals of Surface and Thin Film Analysis,* North-Holland, New York 1986.

J. M. Walls, ed, *Methods of Surface Analysis: Techniques and Applications,* Cambridge University Press, Cambridge, 1989.

R. L. Park and M. G. Lagally, eds., *Solid State Physics: Surfaces,* Academic Press, Orlando, 1985.

R. Mason, C.N.R. Rao, N. Sheppard, M. W. Roberts, and J. M. Thomas, eds., *Studies of the Surfaces of Solids by Electron Spectroscopy,* Royal Society, London, 1986.

H. Ibach, ed., *Electron Spectroscopy for Surface Analysis,* Springer-Verlag, Berlin, 1977.

4. L. J. Clarke, *Surface Crystallography: An Introduction to Low Energy Electron Diffraction,* John Wiley, Chichester, 1985.

 M. A. Van Hove and S. Y. Tong, *Surface Crystallography by LEED,* Springer-Verlag, Berlin, 1979.

 J. B. Pendry, *Low Energy Electron Diffraction,* Academic Press, London, 1974.

 P. M. Marcus and F. Jona, *Determination of Surface Structure by LEED,* Plenum Press, New York, 1980.

5. S. Y. Tong, *Prog. Surface Sci.* **7,** 1 (1975).

6. P. C. Stair, T. J. Kaminska, L. L. Kesmodel, and G. A. Somorjai, *Phys. Rev. B* **11,** 623 (1975).

7. P. C. Wong, M. Y. Zhou, K. C. Hui, and K.A.R. Mitchell, *Surface Sci.* **163,** 172 (1985).

8. F. Jona, *Surface Sci.* **68,** 204 (1977).

9. "Despite its limitations, LEED has delivered well over 50% of all detailed structural analysis so far, in competition with a dozen other structural techniques. LEED has provided about 380 of the 600 detailed surface structural analysis carried out through 1991." M. A. Van Hove, in *Note Book* Vol. 1, J. P. Biberian, ed., R & D International, Marseille, 1993.

10. A. M. Baro, R. Miranda, J. Alaman, N. Garcia, G. Binnig, H. Rohrer, Ch Gerber, and J. L. Carrascosa, *Nature (London)* **315,** 253 (1985).

 P. K. Hansma and J. Tersoff, *J. Appl. Phys.* **61,** R1 (1987).

11. F. M. Hoffmann, *Surface Sci. Rept.* **3,** 107 (1982).

 M. A. Chesters and N. Sheppard, in *Spectroscopy of Surfaces,* R.J.H. Clark and R. E. Hester, eds., John Wiley, New York, 1988.

 Y. J. Chabal, *Surface Sci. Rept.* **8,** 211 (1988).

12. H. Ibach and D. L. Mills, *Electron Energy Loss Spectroscopy and Surface Vibrations,* Academic Press, New York, 1982.

 J. T. Yates, Jr. and T. E. Madey, eds., *Vibrational Spectroscopy of Molecules on Surfaces,* Plenum, New York, 1987.

13. D. Briggs and M. P. Seah, eds., *Practical Surface Analysis by Auger and X-Ray Photoelectron Spectroscopy,* John Wiley, New York, 1983.

14. P. W. Palmberg, G. K. Bohn, and J. C. Tracy, *Appl. Phys. Lett.* **15,** 254 (1969).

15. R. E. Weber and W. T. Peria, *J. Appl. Phys.* **38,** 4355 (1967).

16. T. W. Haas and J. T. Grant, *Appl. Surf. Sci.* **2,** 322 (1979).

 J. T. Grant, *Appl. Surf. Sci.* **13,** 35 (1982).

17. C. R. Brundle and A. D. Baker, eds., *Electron Spectroscopy,* Vols. 1–4, Academic Press, New York, 1977 ff.

D. Briggs and M. P. Seah, eds., *Practical Surface Analysis by Auger and X-Ray Photoelectron Spectroscopy,* John Wiley, New York, 1983.

18. H. H. Brongersma and T. M. Buck, *Nuc. Inst. Meth.* **149,** 1533 (1978).

19. C. M. Greenlief and J. M. White, Secondary ion mass spectrometry, in *Investigations of Surfaces and Interfaces A,* B. W. Rossiter and R. C. Baetzold, eds., Physical Methods in Chemistry Series, 2nd ed., **Vol. IXA,** Wiley Interscience, New York, 1992.

20. T. E. Madey and J. T. Yates, Jr., *J. Vac. Sci. Techn.* **8,** 525 (1971).

    T. E. Madey and R. Stockbauer, in *Methods in Experimental Physics,* R. L. Park, ed., Academic Press, New York, 1986.

    R. D. Ramsier and J. T. Yates, Jr., *Surface Sci. Rept.* **12,** 246 (1990).

21. X. L. Zhou, X. Y. Zhu, and J. M. White, *Surface Sci. Rept.* **13,** 111 (1992).

22. R. B. Hall and S. J. Bares, Pulsed laser induced desorption studies of the kinetics of surface reactions, in *Chemistry and Structure at Interfaces,* R. B. Hall and A. B. Ellis, eds., p. 85, VCH, Weinheim, 1986.

23. P. A. Redhead, *Vacuum* **12,** 203 (1963).

    R. J. Madix, *CRC Crit. Rev. Solid State Mat. Sci.* **7,** 143 (1978).

24. A. M. Bradshaw and K. Scheffler, in *The Chemical Physics of Solid Surfaces and Heterogeneous Catalysis,* **2,** D. A. King and D. P. Woodruff, eds., Elsevier, Amsterdam, 1983.

25. J. C. Riviere, in *Solid State Surface Science,* Vol. I, M. Green, ed., Marcel Dekker, New York, 1969.

26. R. P. Craig and V. Radeka, *Rev. Sci. Instr.* **41,** 258 (1970).

27. A. G. Knapp, *Surface Sci.* **34,** 289 (1973).

28. R. J. Hamers, *Annu. Rev. Phys. Chem.* **40,** 531 (1989).

29. Y. Hasegawa and Ph. Avouris, *Phys. Rev. Lett.* **71,** 1071 (1993).

30. R. J. Hamers, Ph. Avouris, and F. Boszo, *Phys. Rev. Lett.* **59,** 2071 (1987).

CHAPTER

3

Inelastic Scattering and Energy Transfer

3.1 Introduction

The first step in a gas-surface interaction is the collision of the incident gas atom or molecule with the solid surface. This collision can be entirely elastic, with the incident particle leaving the surface after changing direction on collision, and not participating in further heterogeneous reaction. For reaction to subsequently occur, the gas-surface collision must be inelastic, and energy must be transferred between the incident species and the solid surface. If the incident molecule loses enough energy to the surface, it is adsorbed and can then diffuse over the surface and participate in subsequent reaction events. The adsorption, diffusion, reaction, and desorption processes form the subject matter of the later chapters of this monograph. The present chapter is concerned with the detailed study of this initial inelastic collision event. Molecular beam and molecular spectroscopy methods have been applied to this problem, and very detailed information is available about some simple inelastic surface scattering events. This chapter will discuss two specific systems in some detail, to provide examples of the wealth of information that has become available about inelastic scattering from surfaces.

The first studies to consider inelastic scattering of molecules and atoms from solid surfaces probed the energy transfer process by monitoring the angular distribution of the particles scattered from the surface as a function of incident translational energy and angle. Observations of subspecular and supraspecular angular distributions[1] were the first direct indications that the incident gas phase species could exchange energy with the surface. These measurements were ini-

25

tially carried out with rare gas atom scattering from characterized surfaces,[2] and were then extended to molecular scattering from clean and adsorbate covered surfaces.[3] The phonon excitation or absorption reflected in the sub- and supraspecular, rather broad distributions lead to the development of the application of time-of-flight (TOF) methods to the study of inelastic processes. Using a well-collimated and monoenergetic incident beam, and long flight path TOF measurements, single phonon creation and annihilation events could be observed,[4] as well as the multiple phonon excitation that resulted in the broadened angular distributions observed for incident beams with a broad energy distribution. Detailed studies of the phonon spectra of the surfaces of well-characterized solids developed from these measurements.[5] This subject has been reviewed extensively,[6] and is tangential to the subsequent heterogeneous reaction processes with which we are concerned, so it will not be discussed further.

These early angular distribution measurements also suggested the possibility of internal energy excitation in the scattered molecule, as well as energy transfer between the translational energy of the beam and the phonon modes of the surface. This internal excitation was reflected in distinct differences among the angular distributions of scattered hydrogen, deuterium, and hydrogen deuteride.[7] The observed broadening in the case of D_2 and HD compared to the narrow specular distribution for H_2 scattering from Ag(111) suggested rotational excitation of the scattered D_2 and HD molecules. For D_2 and HD the first rotational level is low enough in energy that sufficient collision energy was available to result in rotational excitation, where this would not be the case for H_2. This possibility of $T \rightarrow R$ energy transfer on collision has been extensively studied in the following years, using state-selective probes of the scattered molecule to determine the rotational excitation as a function of the gas surface collision parameters. The two examples of this chapter discuss this topic in detail.

There have also been some studies concentrating on the vibrational and electronic excitation of molecules colliding with surfaces.[8] As vibrational and electronic energy levels are in general much higher than rotational levels, this excitation process is much less likely at the energies typical of thermal gas-surface collisions. However, at higher collision energies these processes can occur, and several groups have begun to examine the dynamics of these processes in detail. These hyperthermal collision events can involve ion-surface collisions, and can have importance for technologically interesting processes such as ion etching,[9] ion-assisted thin film growth,[10] and surface-induced dissociation in analytical mass spectrometry.[11] The examples discussed in this chapter will not in general deal with these higher energy collisions, but will concentrate rather on the energy transfer upon gas-surface collision at thermal energy.

3.2 Experimental Methods

The examples discussed in this chapter utilize molecular beam scattering methods to control the gas-surface collision parameters. These methods figure strongly in a number of the examples here, and particular aspects of molecular beam scattering have been described in detail both earlier and later in this monograph. The important point to note here is that the well-collimated nozzle beam source provides a beam of molecules or atoms with a well-defined incident geometry, and a well-defined monoenergetic incident translational energy. This translational energy can be readily changed by changing the temperature of the nozzle source, or by seeding the molecule of interest in a heavier or lighter carrier gas. The details of source design and the qualities of these molecular beam sources have been covered extensively in the literature.[12]

As pointed out in Chapter 2, it is essential that the surface under study be well characterized as to structure and composition. The methods described in Chapter 2 allow this characterization, and when the well-characterized surface is mounted in an ultrahigh vacuum scattering chamber, and the well-characterized molecular beam source is incident on this surface, a great deal of information can be obtained about the inelastic scattering process. In addition to the measurement of the angular distribution of the scattered species, TOF methods can be used to measure the scattered molecule velocity. Direct measurement of the internal state distribution of the scattered molecule is also possible now, using a number of laser and electron-based techniques. Again, these methods figure very prominently in several of the other examples described in later chapters, and detailed description of some of them appear in those chapters. One particular method that is important to the first example of this chapter, laser-induced fluorescence, is described briefly here.

Laser-induced fluorescence (LIF) is a relatively general, highly sensitive state selective molecular detection technique.[13] It measures indirectly the absorption spectrum of a molecule, and in favorable cases can directly give the population distribution for particular rotational–vibrational levels of the molecule. The absorption spectrum is measured by excitation of the molecule to an electronically excited state that fluoresces. A narrow line tunable laser source is used to scan across an electronic absorption band of the molecule of interest. When the laser is in resonance with a particular internal energy transition in this electronic band, a fraction of the molecules in the ground state of that transition will be excited to the electronically excited state and fluoresce. The intensity of the fluorescence is proportional to the population of the molecule in that particular internal energy state. As the laser is scanned across the band, this excitation spectrum obtained by collecting the fluorescence radiation, directly reflects the population distribution in the internal energy levels of the detected molecule. The radiative behavior of the excited state must be well known for this approach to be quantitative. Fortunately this is the case for a number of interesting diatomic and triatomic molecules. Larger molecules generally do not fluoresce efficiently upon excitation, and so the LIF method is not applicable to their investigation.

Nitric oxide, NO, is particularly well suited to probing by laser induced fluorescence,[14] and the first detailed studies of gas surface translational to rotational energy transfer focussed on the scattering of NO from various surfaces. Excitation of the $^2\Sigma \leftarrow {}^2\Pi$ transition in NO can be accomplished with radiation tunable near 225 nm. This wavelength range is readily available in high intensity, and the NO emission spectrum is well known, so LIF detection of the internal energy of this molecule is fairly straightforward. In addition, the $^2\Pi$ ground state is spin-orbit split, with a level spacing of about 120 cm^{-1}, so information about the interaction of the spin-orbit states with the scattering surface can also be obtained.[15]

3.3 Illustrative Examples

3.3.1 Inelastic Scattering of NO from Ag(111)

The first inelastic scattering system to be extensively and directly studied using state sensitive detection of the scattered molecules was the scattering of NO from the Ag(111) surface. Two groups have obtained an extensive range of information about this system. In these cases, a supersonic beam expansion of NO seeded in helium was used as the incident beam, and the Ag(111) surface was cleaned by ion bombardment and characterized using LEED and Auger electron spectroscopy under ultrahigh vacuum conditions. Kleyn et al.[16] concentrated on the incident beam energy range from about 0.2 to 1.2 eV, and Kubiak et al.[17] studied the incident energy range from 0.05 to 0.3 eV. In both studies the expansion conditions were such that the rotational temperature of the incident beam was in the range of 5–50 K. Both the angle of incidence of the beam and the angle of detection of the scattered molecules could be varied.

The scattered molecules were detected by laser-induced fluorescence using light in the range of 225 nm to scan the $^2\Sigma \leftarrow {}^2\Pi$ transition of the NO. The polarization of the laser light with respect to the surface normal could be varied. Internal state distributions were obtained by scanning the laser across this transition for particular conditions of incident angle, energy, surface temperature, and detector angle. The internal energy data is presented as Boltzmann plots as shown in Figure 3.1. The ln of the line intensity divided by $(2J + 1)$ is plotted versus the total internal energy of the molecule. If the population distribution can be described as a Boltzmann distribution at a particular temperature, such a plot yields a straight line whose slope provides the rotational temperature. As can be seen from the results of Figure 3.1, these distributions are not well described by a fit to a single straight line. At low internal energies, the data do fit a line, but there is clearly excess population in the rotational levels at higher **J**. This effect is described as a rotational rainbow.

This effect is called a ''rainbow'' in analogy with the terminology used in gas phase scattering for the peak in scattering intensity ascribed to the contribution of many individual scattering trajectories to a single final scattering angle or narrow range of angles.[18] In the case of molecular scattering from solid surfaces, a number of possible molecular surface collision orientations can contribute to a range of rotational excitations for the scattered molecule. This might

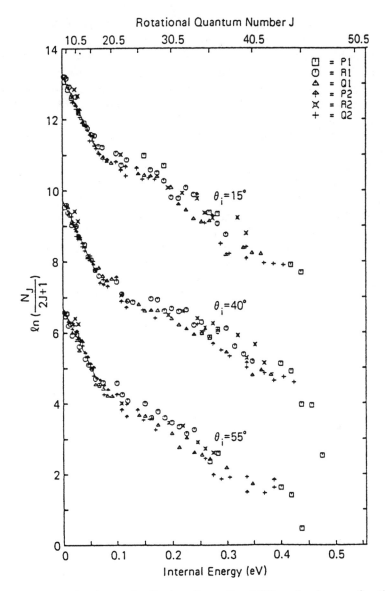

Figure 3.1. Rotational state distributions for scattered NO molecules as a function of internal energy $E_{int} = E_{rot} + E_{SO}$. E_{SO} is the spin-orbit energy, which is zero for P1 and R1 branch transitions. A **J** scale is given (top). For $\theta_i = 15°$ and $\theta_i = 40°$, $E_n = 0.44$ eV. For $\theta_i = 55°$, $E_n = 0.37$ eV. In all cases $T_S = 650$ K. T_R as determined from a fit of the data (full line) at low **J** is 360 ± 15 K for all fits. After ref. 23.

best be visualized by considering the collision of an initially nonrotating molecule with a flat, noncorrugated surface. Figure 3.2 indicates the geometry in question, and helps to illustrate this concept. The angle between the molecular axis and the surface normal is defined as γ. Molecule-surface collisions with $\gamma = 0$ or $\gamma = \pi$ do not result in rotational excitation of the scattered molecule. Similarly, collisions with $\gamma = \pi/2$ will not excite rotational motion. For collision geometries with γ between zero and $\pi/2$, the rotational excitation must go through a maximum, as indicated in Figure 3.2. This maximum in excitation as a function of the collision geometry gives rise to the rotational rainbow effect, the additional population at higher rotational levels than would be expected from a Boltzmann distribution.[19] Model calculations of a rigid rotor scattering from a static flat surface, using a simple symmetrical Morse potential interaction,[20] do a good job of reproducing the qualitative shape of the plots of level population as a function of rotational energy, as shown in Figure 3.3. While the purely classical trajectory calculations suggest the presence of a rotational rainbow, the quantum mechanical calculations do a much better job of describing the shape of the experimental data. In particular, when the distribution of rotational energies of the incident molecule is included, as in the plot of Figure 3.3d, the agreement with the experimental data is quite good.

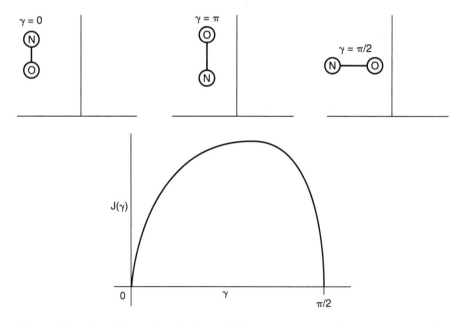

Figure 3.2. Description of molecular collision geometry. γ is the angle between the molecular axis and the surface normal. $J(\gamma)$ is the excitation function, which must go through a maximum between 0 and $\pi/2$ (and between $\pi/2$ and π for a heteronuclear diatomic such as NO).

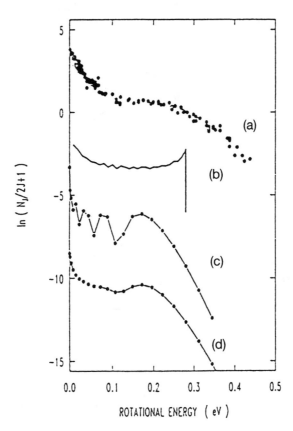

Figure 3.3. Boltzmann plot of flux as function of rotational energy. The vertical scales are shifted for convenience. (a) Experimental data from ref. 16; (b) results of classical computations for incident energy 0.65 eV and $T_R = 0$, (c) results of quantum-mechanical calculations for the same parameters as in (b), (d) results of quantum-mechanical calculations for the same parameters as in (b) except that T_R is 30 K. After ref. 20.

Even for the best case however, the straight line behavior observed experimentally at low rotational energies is not reproduced. Significant curvature in the calculations for low rotational excitation is observed, as can be seen in Figure 3.3. The observation of two distinct regions of behavior as a function of rotational energy suggested to Voges and Schinke that perhaps the experimental data described here could be interpreted in terms of a strongly asymmetric potential of interaction for the rigid rotor scattering from the static flat surface.[21] This treats the NO molecule as a heteronuclear diatomic, and results in a theoretical description comprised of two rotational rainbows. The strong asymmetry in the interaction potential results in two extrema in a plot of $J(\gamma)$ vs γ, as illustrated in Figure 3.4. Physically, this implies that the interaction of the N-end of the molecule with the surface is significantly different than the interaction of the

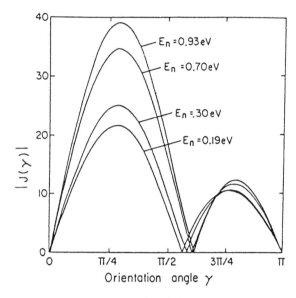

Figure 3.4. Classical excitation function $|J(\gamma)|$ versus orientation angle γ for an asymmetric potential. After ref. 21.

O-end of the molecule with the surface. This treatment of the scattering gives good agreement with the experimental data across the range of observed final rotational energies. In particular, the straight line behavior at low rotational energies is reproduced, and the excess rotational population at high rotational energies is seen. These results clearly indicate that the NO/Ag(111) scattering potential is strongly asymmetric, and that the orientation of the molecule on approach to the surface affects the scattering dynamics.

Laser-induced fluorescence can also be used to probe the rotational polarization of the scattered molecule in the NO/Ag(111) scattering system.[22] The orientation of the final angular momentum vector is obtained by observing the dependence of the LIF intensity on the direction of linear polarization of the excitation radiation. The final angular momentum vector is found to be strongly oriented perpendicular to the surface normal, suggesting that molecule–surface collisions result in the excitation of cartwheeling motions of the scattered NO molecule. Helicoptering rotational motion, with the final angular momentum vector parallel to the surface normal, is not observed. There is also a strong dependence of the polarization on the final J state of the scattered molecule. In the low **J** region, where the population intensity is well fit by a Boltzmann distribution, no polarization of the final angular momentum vector is observed. In the high **J** region, where a rotational rainbow is observed, due to the direct inelastic nature of the molecule–surface scattering, a strong polarization of the final angular momentum vector is seen. Trajectory calculations suggest that this difference may be due to differences in the actual initial orientation of the molecule with respect to the surface prior to collision.

The direct inelastic nature of the scattering is also indicated in rotationally state resolved angular distribution measurements carried out on the NO/Ag(111) scattering system.[23] The results of such measurements are illustrated in Figure 3.5, where it is seen that the angular distribution for rotational states with $J \leq$ 23.5 is clearly specular. For higher J states, $J = 40.5$ for example, the angular distribution is shifted away from the normal, above specular. Both of these results suggest that the scattering is direct inelastic in nature, and does not involve trapping of the NO molecule in the surface adsorption well.

Detailed measurements of the velocity distribution of the scattered NO molecules, resolved by final rotational state, suggest that there is a correlation between the final velocity and the final rotational energy of the scattered NO molecule.[24] In particular, the scattered NO molecule velocity is found to decrease with increasing rotational excitation for fixed incident translational energy and angle. This efficient transfer of incident translational energy to both phonons in the surface and rotational energy of the scattered molecule appears to limit the total translational energy loss in gas-surface collisions, at least for relatively high collision energies. Trajectory calculational modeling of these results, and the consideration of a simple kinematic model of the collision, leads to the suggestion that correlation between the final velocity and the final rotational energy of the scattered molecule is associated with the orientation of the molecule when it collides with the surface. Those collision geometries leading to high rotational excitation, that is those with γ close to zero (O-end down), result in less phonon excitation on collision. Collision geometries consistent with less rotational excitation, that is with the N-end down ($\gamma = \pi$), or with the molecular axis close to

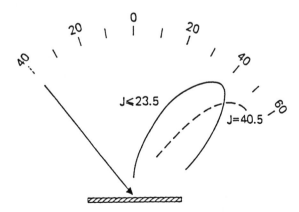

Figure 3.5. Angular distribution of NO molecules scattered from Ag(111) as detected using LIF. The solid line is a composite of three angular scans with $J \leq 23.5$ and $E_i =$ 1.1 eV. The dashed line shows an angular distribution for $J = 40.5$ and $E_i = 0.75$ eV. In all cases $T_S = 650$ K. Q1 transitions have been used for all J. After ref. 23.

parallel to the surface ($\gamma = \pi/2$), result in more phonon excitation upon collision. Agreement between model calculations and the experimental data at the highest collision energies and for the lowest rotational excitation is less satisfactory than that for higher final rotational states.

Another very interesting aspect of NO scattering from the Ag(111) surface that has been experimentally explored is the possible differences in scattering that may be observed due to the small spin-orbit splitting of the NO molecule. These two states, the $^2\Pi_{\frac{1}{2}}$ and $^2\Pi_{\frac{3}{2}}$ are separated by 123 cm^{-1} in the lowest rotational state, and result in the observation of well-separated rotational sub-bands in the molecule.[25] Comparison of the LIF intensities of the two bands then provides information about the partitioning of energy into the two electronic states. The two spin-orbit states differ in the orientation of the electron density of the unpaired electron in the π-orbital of the NO molecule relative to the molecular rotation axis **J**. In one case the electron density in the π orbital is parallel to **J**, in the other it is perpendicular to **J**. High translational energy collisions of NO with the Ag(111) surface result in nonequilibrium populations of the spin-orbit states of NO.[26] In particular, for collisions resulting in significant rotational excitation, there is strong mixing of the spin-orbit states on collision, with a definite preference for the population of the spin orbit state with the π-electron density perpendicular to **J**. This observation suggests that rotational excitation is enhanced for NO collisions with the singly occupied π-orbital facing the surface. For scattering resulting in low rotational excitation, the spin-orbit population is essentially in equilibrium with the rotational state distribution. The ratio of spin-orbit temperature to measured rotational temperature in this region is equal to one, within the error of the experiment, even though neither population is in equilibrium with the surface temperature. At lower collision energies, preferential population of spin-orbit levels is not observed, over the range of final rotational states up to **J** = 40.5. In this case, no effect due to the different orientation of the π-electron density in the different spin-orbit states was observed for these lower translational energy collisions.

More recent experiments have been able to directly probe the asymmetric nature of the NO/Ag(111) interaction hinted at in the earlier experiments described above. Using hexapole focusing fields, Kleyn and co-workers have been able to prepare a high purity state selected beam of NO molecules, and to investigate directly the effect of molecule orientation on the scattering of NO from the Ag(111) surface.[28] The NO beam, formed in a pulsed nozzle expansion and selected in the $^2\Pi_{\frac{1}{2}}$ ($J = \frac{1}{2}$, $\Omega = \frac{1}{2}$, $M_J = \frac{1}{2}$) state by the hexapolar field is incident on the scattering surface in the presence of a strong electrostatic guiding field normal to the surface, provided by a rod positioned above and parallel to the surface. This field orients the state selected beam of molecules so that collisions occur with either the N-end down (positive rod voltage) or O-end down (negative rod voltage). The scattered NO molecules were then detected mass spectrometrically to measure the scattered angular distribution as a function of incident molecule orientation,[29] or using multiphoton ionization to probe the final rotational state of the scattered oriented molecules.[30]

In these studies, steric effects were observed in both the angular distributions and in the rotational state distributions of the scattered molecules. For scattering with the N-end of the molecule first, the angular distribution is peaked below the specular angle (toward the surface normal). For O-end scattering, the angular distribution is shifted toward the surface tangent, as compared to the angular distribution for randomly oriented incident molecules.[29] Given the flat scattering surface, rotational excitation must result from changes in the normal component of the incident molecule velocity, which is then reflected in shifts in the angular distribution maximum. These observations suggest that O-end scattering results in more rotational excitation than N-end scattering. This conclusion is supported by the direct measurement of the rotational state of the scattered molecules as a function of the incident molecule orientation.[30] Low J states are preferentially populated by N-end collisions, while the higher $J = 18.5$ state is preferentially populated by O-end collisions. These direct observations are fully consistent with the asymmetric potential model of Voges and Schinke,[21] and with the earlier angular distribution and TOF measurements described above.

A number of very detailed experimental studies of the rotationally inelastic scattering of NO from the Ag(111) surface have been carried out. They suggest that this system is controlled by direct inelastic scattering of the molecular species from the repulsive wall of the gas-surface interaction potential, and that this interaction potential is highly asymmetric. Trajectory calculations, both classical and quantal, have been used to model this system and to derive detailed information about the form of the interaction potential. This system serves as an excellent example of the level of information about heterogeneous dynamic processes that is available from state-of-the-art experiment coupled with theoretical calculation. In our discussion we have concentrated on the rotationally inelastic scattering process, and its coupling to final velocities and spin-orbit populations. Other investigators have also examined vibrationally inelastic scattering of NO from various surfaces,[31] including the study of the state selected scattering of vibrationally excited NO from the silver surface.[32] The rich molecular spectroscopy of the NO molecule provides an experimental handle that allows one to obtain a great deal of information about the interaction potential governing the scattering of this molecule from well-characterized solid surfaces.

3.3.2 Direct Inelastic Scattering of Nitrogen from Ag(111)

A very detailed state selective study of the scattering of molecular nitrogen from the Ag(111) surface has been carried out by Sitz et al.,[33–37]. This study relies on the development of an efficient scheme for the (2 + 2) REMPI detection of nitrogen internal state distributions.[38] This scheme uses tunable radiation from a frequency doubled YAG pumped dye laser to excite the (1,0) band of a two photon transition in N_2. This transition, the $a^1\Pi_g \leftarrow X^1\Sigma_g^+$, lies in the 283–285 nm range in the near UV. The same laser light is used to ionize the excited molecules by a further 2 photon absorption process. Pulse energies in the range of 15 mJ per pulse at 283 nm were sufficient to detect individual rotational levels of nitrogen at a nitrogen density equivalent to about 10^{-7} torr.

Data typical of this REMPI detection scheme for the incident nitrogen beam and the beam scattered from the Ag(111) surface are shown in Figure 3.6. The clear difference between the low rotational excitation of the incident beam and the significant excitation of the scattered beam into high rotational levels is apparent, as is the excellent signal strength and signal-to-noise ratio for the scattered beam data. Also evident is the alternation in intensity of the rotational lines due to the *ortho* and *para* nuclear spin states of the spin $= 1$ ^{14}N nucleus. In the scattering experiments, the detection laser is focused above the scattering surface between the grids of a TOF mass spectrometer. The laser beam propagates perpendicular to the plane described by the surface normal and the incident molecular beam, and can be rotated to probe different final scattering angles with an angular resolution of about 10°. The detection laser is linearly polarized, and the direction of polarization with respect to the surface normal can be varied. The ions formed by the excitation–ionization process are detected by the TOF mass spectrometer.

As pointed out above, the scattered beam data of Figure 3.6 shows significant excitation to high rotational levels. In addition, there are different intensities seen for transitions in different spectral branches originating from the same

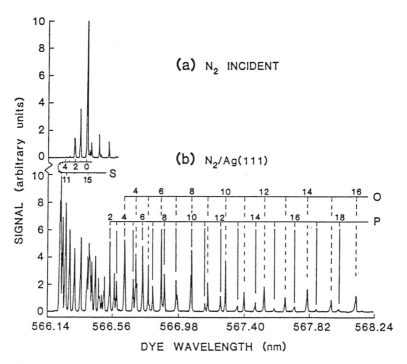

Figure 3.6. REMPI spectra of (a) the incident N_2 beam and (b) the scattered N_2 beam. Conditions were in (a) nozzle orifice $d = 400$ μm, stagnation pressure $P_0 = 70$ psig, seeding ratio of 22% N_2 and 78% H_2, and in (b) $\theta_i = 15°$, $\theta_f = 20°$, and $T_s = 90K$. Line assignments are made based on tabulated energy levels. After ref. 34.

ground state level, when compared to the spectrum of the isotropic background gas. This suggests that the scattering results in polarization of the scattered molecules. This polarization is described by the multipole moments of the angular momentum distribution. Sitz et al. analyze their results in terms of the multipole moments of the angular momentum distribution, and discuss the scattering process in terms of the rotational excitation,[34] the angular momentum alignment,[34] and the angular momentum orientation[35] for the scattered nitrogen. The degree of polarization of an individual rotational line is obtained by recording the intensity of that line while changing the direction of polarization of the laser with respect to the surface normal. Data typical of this sort of measurement are shown in Figure 3.7 for two separate branch J = 15 rotational levels.[34] This clear variation in intensity with laser polarization and spectral branch indicate that the

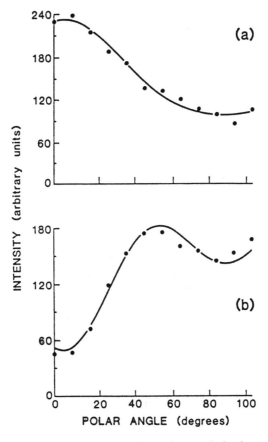

Figure 3.7. REMPI integrated line intensities vs laser polarization angle for (a) O(15) and (b) for P(15). The laser polarization angle is measured with respect to the surface normal. The symbols are the measured data points. After ref. 34.

polarization of the scattered sample must be understood prior to the extraction of rotational state distribution information.

This analysis, which is carried out and described in detail in references 34–36, reaches the conclusion that the scattered N_2 is highly aligned, with the rotational vector **J** lying in a plane parallel to the surface. For intermediate values of **J**, the alignment is less than would be predicted for rigid rotor scattering from a flat surface. At high values of **J**, the alignment approaches the limiting value for the **J** vector to be perpendicular to the surface normal.[34] Once the polarization and degree of alignment in the scattered data are understood, then the rotational population distributions can be derived from the data.

A typical Boltzmann plot of the scattered data is shown in Figure 3.8. Again, as was seen for the scattering of NO from the Ag(111) surface, there is clearly excess population at high **J** values, precluding the description of the rotational state distribution as a Boltzmann distribution with a characteristic temperature. The rotational rainbow effect described earlier is again seen, and the prominence

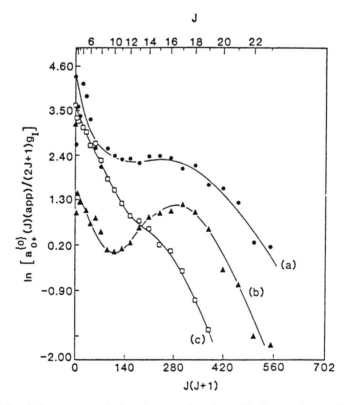

Figure 3.8. Boltzmann population plots as a function of final scattering angle for $\theta_1 = 30°$, $E_1 = 0.3$ eV, and $T_s = 90$ K. Final angles are (a) $\theta_f = 35°$, (b) $\theta_f = 50°$, and (c) $\theta_f = 25°$. After ref. 34.

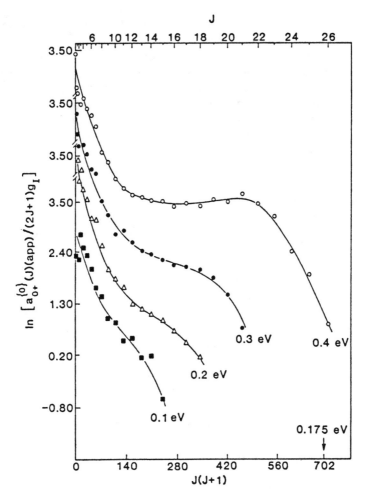

Figure 3.9. Boltzmann population plots as a function of incident beam energy for θ_i = 15°, θ_f = 20°, and T_s = 90 K. After ref. 34.

of the rainbow varies with final scattering angle. Figure 3.9 illustrates the Boltzmann plots of the scattered beam data as a function of incident beam energy, and also shows the strong dependence of the rainbow effect on increasing incident energy.

Time-of-flight measurements of the translational energy of the scattered N_2 molecules as a function of final rotational energy indicate that the velocity of the scattered molecules decreases with increasing rotational excitation.[34] However, the sum of rotational and translational energy of the scattered molecules increases with increasing rotational energy, suggesting that the amount of energy transferred to phonons on collision decreases with increasing rotational energy.

As with the scattering of NO from the Ag(111) surface, there seems to be a limit to the total amount of incident translational energy lost on collision.

The detailed rotational, alignment, and translational energy information provided by these experiments provides a set of limits for possible potentials describing the N_2/Ag(111) interaction. As with the scattering of NO from the Ag(111) surface at higher energies, the scattering interaction here is clearly a direct inelastic process. The strong dependence of rotational excitation on incident energy and final scattering angle and the relative insensitivity to the surface temperature all indicate that the scattering is probing the N_2/Ag(111) interaction potential directly. The alignment of the scattered molecules is preferentially with **J** in the plane of the surface. At high final rotational energies, **J** approaches the limit of being perpendicular to the surface normal.

Further limits on the form of the interaction potential can be obtained by exploring the orientation of the rotationally excited scattered molecules.[35] Orientation here refers to the clockwise or counterclockwise sense of the rotation of the scattered molecule. This quality of the scattered rotational state distribution is probed by using elliptically polarized light in the REMPI detection scheme. The degree of elliptical polarization is designated by β, where $\beta = 0$ corresponds to linear polarization, and when $\beta = \pm 45°$, the light is left or right circularly polarized. For intermediate values, the laser probe light is said to be elliptically polarized. Making REMPI measurements of the scattered molecules as a function of β corresponds to changing the sense of rotation of the **E** field with respect to the laser propagation direction. This should make the REMPI signal sensitive to the sense of rotation of the molecule in a plane perpendicular to the laser propagation direction. Figure 3.10 shows typical orientation data for two particular rotational lines for N_2 scattering from the Ag(111) surface. There is clearly an asymmetry in the data, suggesting a preferential handedness to the rotation of the scattered molecule. The degree and sign of the alignment were found to vary quite strongly with final scattering angle and final rotational state. This observation suggests that the flat surface-hard ellipsoid model of the scattering that successfully reproduced the direct inelastic scattering of NO from the Ag(111) surface is not likely to reproduce the data obtained here. The fact that the direction of rotation of the scattered N_2 molecule depends on the geometry of the collision, as reflected in the final scattering angle, and in the degree of rotational excitation, suggests that there must be forces in the plane of scattering operating to control the direction of rotation during the collision.

The inclusion of forces in the surface plane is realized in classical trajectory modeling of this system by incorporation of frictional forces in the plane of the surface. A frictional hard-cube hard-ellipsoid model was developed[35] that very successfully reproduces the rather complicated orientation, alignment, and population distribution measurements for this system. This added friction represents the laterally averaged forces in the plane of the surface, and reproduces nicely the experimental observations, including the change in sign of the orientation with changing final rotational state and scattering angle. The physical origin of the forces modeled by this frictional term is the weak corrugation of the Ag(111)

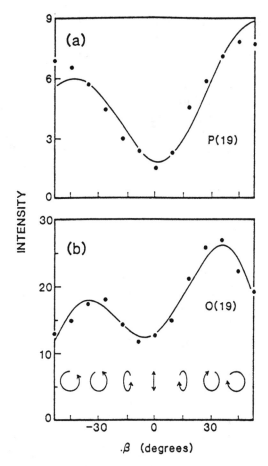

Figure 3.10. Orientation data and fits for (a) P(19) and (b) O(19) for N_2 scattered off Ag(111). Conditions were $\theta_i = 30°$, $\theta_f = 35°$, $E_i = 0.3$ eV, and $T_s = 90$ K. The curves are asymmetric about $\beta = 0°$ indicating that the $\mathbf{J} = 19$ level is oriented. After ref. 35.

surface. A rather strong orientation effect is experimentally observed for this very smooth (weakly corrugated) surface, suggesting that detailed measurements of the orientation and alignment of scattered molecules provides a very sensitive probe of the in-plane and vertical details of the gas-surface interaction potential.

Further studies of this system, restricting the measurements to incident trajectories along the surface normal,[36] provide information that takes advantage of the cylindrical symmetry of this scattering geometry. Molecules probed along the backscattered (normal) direction exhibit no angular momentum orientation. Those probed away from the surface normal, however, show strongly varying degrees of rotational excitation and angular momentum orientation. The degree of orientation is seen to depend strongly on final rotational state, and also to depend on the final scattered angle. Stochastic trajectory calculations utilizing a realistic empirical interaction potential were carried out to simulate these observations.[36,37] A study of the individual trajectories suggests that the final rotational state and degree of orientation are determined primarily by the initial molecular orientation of the molecule, that is, the angle of the intermolecular axis with respect to the surface normal upon collision. The final scattering angle, on the other hand, is determined principally by the actual two-dimensional impact parameter of the molecule within the surface unit cell. These sorts of detailed measurements, in connection with realistic trajectory calculations, are able to differentiate between the parts of the gas-surface interaction potential that control the dynamics of the surface scattering process.

3.4 Comments

The examples presented in this chapter, perhaps more than any of the other case studies discussed in this book, illustrate two important points about the study of heterogeneous reaction dynamics. The first point is that these examples provide a taste of the overwhelming amount of detail about heterogeneous interactions that is available for favorable systems. The molecular spectroscopy of both NO and N_2 is well enough understood that powerful techniques are available to examine every detail of the molecule-surface interaction. Our detailed knowledge of the spectroscopy of these molecules allows the use of LIF and REMPI probes that provide internal state distribution information. Application of these methods to other molecules of interest is continually occurring, and as molecular spectroscopic probes are developed they will certainly be applied to other heterogeneous systems of interest. The examples described here concentrate on the use of direct inelastic scattering as a probe of the gas-surface interaction potential, but these probes have been and will continue to be used in the context of adsorption, desorption, and reaction on surfaces, allowing the construction of the interaction potentials governing these important heterogeneous processes as well.

The second point, which is clearly illustrated by the examples presented here, is the essential interaction between high quality experiment and theoretical description. The detailed rotational, orientation, and alignment data presented here could not have been physically connected to the gas-surface scattering event without the use of trajectory calculations and the development of simple kinematic models to describe the scattering process. This interplay between experiment and theory is very clear in these examples, and enables the construction of the gas-surface interaction potential for these systems. Exactly the same sort of detailed interplay is essential for the development of an understanding of the interaction potentials that govern reactions at surfaces. This is where the future of heterogeneous reaction dynamics research lies.

References

1. J. N. Smith, Jr., H. Saltsburg, and R. L. Palmer, *J. Chem. Phys.* **49**, 1287 (1968).

2. R. Sau and R. P. Merrill, *Surface Sci.* **34**, 268 (1973).

3. S. L. Bernasek and G. A. Somorjai, *J. Chem. Phys.* **60**, 4552 (1974).

4. M. N. Bishara and S. S. Fisher, *J. Chem. Phys.* **52**, 5661 (1970).

5. B. R. Williams, *J. Chem. Phys.* **55**, 1315 (1971).

6. J. P. Toennies, in *Surface Phonons,* W. Kress and F. W. deWitte, eds., Springer Verlag, Berlin, 1991, p. 111.

7. R. L. Palmer, H. Saltsburg, and J. N. Smith, Jr., *J. Chem. Phys.* **50**, 4661 (1969).

8. M. Asscher, W. L. Guthrie, T.-H. Lin, and G. A. Somorjai, *J. Chem. Phys.* **78**, 6992 (1983); *Phys. Rev. Lett.* **49**, 76 (1982); J. Misewich, H. Zacharias, and M.M.T. Loy, *J. Vac. Sci. Technol. B* **3**, 1474 (1985); A. Amirav, *Comments At. Mol. Phys.* **24**, 187 (1990).

9. D. J. Thomson and C. R. Helms, *Surface Sci.* **236**, 41 (1990).

10. S. R. Kasi, H. Kang, C. S. Sass and J. W. Rabalais, *Surface Sci. Rep.* **10**, 1 (1989).

11. R. G. Cooks, T. Ast, and M. A. Mabud, *Int. J. Mass Spectrom. Ion Proc.* **100**, 209 (1990).

12. G. Scoles, ed., *Atomic and Molecular Beam Methods,* Oxford University Press, New York, Vol. 1 (1988), Vol. 2 (1992).

13. J. L. Kinsey, *Annu. Rev. Phys. Chem.* **28**, 349 (1977).

14. R. N. Zare and P. J. Dagdigian, *Science* **185**, 739 (1974).

15. M. P. Sinha, C. D. Caldwell, and R. N. Zare, *J. Chem. Phys.* **61**, 491 (1974).

16. A. W. Kleyn, A. C. Luntz, and D. J. Auerbach, *Phys. Rev. Lett.* **47**, 1169 (1981).

17. G. M. McClelland, G. D. Kubiak, H. G. Rennagel, and R. N. Zare, *Phys. Rev. Lett.* **46**, 831 (1981).

18. R. Schinke, *Chem. Phys.* **34**, 65 (1978); W. Schepper, U. Ross, and D. Beck, *Z. Phys.* **A290**, 131 (1979); K. Bergman, U. Hefter, and J. Witt, *J. Chem. Phys.* **72**, 4777 (1980); J. A. Serri, C. H. Becker, M. B. Elbel, J. L. Kinsey, W. P. Moskowitz, and D. E. Pritchard, *J. Chem. Phys.* **74**, 5116 (1981).

19. J. A. Barker and D. J. Auerbach, *Surface Sci. Rep.* **4**, 1 (1984).

20. J. A. Barker, A. W. Kleyn, and D. J. Auerbach, *Chem. Phys. Lett.* **97**, 9 (1983).

21. H. Voges and R. Schinke, *Chem. Phys. Lett.* **100**, 245 (1983).

22. A. C. Luntz, A. W. Kleyn, and D. J. Auerbach, *Phys. Rev. B* **25**, 4273 (1982).

23. A. W. Kleyn, A. C. Luntz, and D. J. Auerbach, *Surface Sci.* **117**, 33 (1982).

24. J. Kimman, C. T. Rettner, D. J. Auerbach, J. A. Barker, and J. C. Tully, *Phys. Rev. Lett.* **57**, 2053 (1986).

25. G. Herzberg, *Spectra of Diatomic Molecules,* Van Nostrand, Princeton, 1950.

26. A. C. Luntz, A. W. Kleyn, and D. J. Auerbach, *J. Chem. Phys.* **76**, 737 (1982).

27. G. D. Kubiak, J. E. Hurst, Jr., H. G. Rennagel, G. M. McClelland, and R. N. Zare, *J. Chem. Phys.* **79**, 5163 (1983).

28. M. G. Tenner, E. W. Kuppers, W. Y. Langhout, A. W. Kleyn, G. Nicolasen, and S. Stolte, *Surface Sci.* **236,** 151 (1990).

29. E. W. Kuipers, M. G. Tenner, A. W. Kleyn, and S. Stolte, *Nature (London)* **334,** 420 (1988).

30. M. G. Tenner, F. H. Geuzebroek, E. W. Kuipers, A. E. Wiskerke, A. W. Kleyn, S. Stolte, and A. Namiki, *Chem. Phys. Lett.* **168,** 45 (1990).

31. C. T. Rettner, J. Kimman, F. Fabre, D. J. Auerbach, and H. Morawitz, *Surface Sci.* **192,** 107 (1987). C. T. Rettner, F. Fabre, J. Kimman, and D. J. Auerbach, *Phys. Rev. Lett.* **55,** 1904 (1985).

32. J. Misewich and M.M.T. Loy, *J. Chem. Phys.* **84,** 1939 (1986).

33. G. O. Sitz, A. C. Kummel, and R. N. Zare, *J. Vac. Sci. Technol. A* **5,** (1987).

34. G. O. Sitz, A. C. Kummel, and R. N. Zare, *J. Chem. Phys.* **89,** 2558 (1988).

35. G. O. Sitz, A. C. Kummel, R. N. Zare, and J. C. Tully, *J. Chem. Phys.* **89,** 2572 (1988).

36. A. C. Kummel, G. O. Sitz, R. N. Zare, and J. C. Tully *J. Chem. Phys.* **89,** 6947 (1988).

37. A. C. Kummel, G. O. Sitz, R. N. Zare, and J. C. Tully, *J. Chem. Phys.* **91,** 5793 (1989).

38. K. L. Carleton, S. R. Leone, and K. H. Welge, *Chem. Phys. Lett.* **115,** 492 (1985).

CHAPTER

4

Adsorption, Epitaxial Growth, and Adsorbate Interactions

The next step in the gas-surface interaction is the transfer of sufficient energy in the gas-surface encounter to result in trapping and subsequent adsorption of the incident atomic or molecular species. This adsorption process is preliminary to any surface reaction, or to the incorporation of the adsorbate in a layer on the surface. This chapter considers, through illustrative examples, the fate of a surface trapped species. Its subsequent adsorption, adsorbate–adsorbate interactions preliminary to surface reaction, and incorporation into a growing epitaxial film are all discussed in the examples comprising this chapter. Thermal energy atom scattering (TEAS) is used to probe the dynamics of the adsorption and growth process.

4.1 Experimental Methods

A great deal of information about the dynamics of the adsorption process can be obtained by the use of TEAS from the surface. TEAS, a method pioneered by Poelsema and Comsa,[1] monitors the intensity of a well-collimated, specularly reflected helium atom beam from the surface as it is exposed to adsorbate species. For clean, atomically smooth, close packed surfaces such as the Pt(111) surface, the He beam specular reflectivity is very high. The surface behaves essentially as a perfect mirror for the He beam. Adsorption of an atomic or molecular adsorbate results in considerable diffuse scattering of the He beam, and a sharp reduction in specular intensity as adsorbate coverage increases. Figure 4.1 illustrates schematically an apparatus for TEAS studies of adsorption.

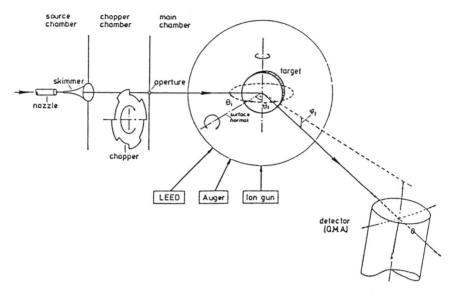

Figure 4.1. Schematic of thermal energy atom scattering apparatus. After ref. 1.

For a perfectly diffuse scatterer randomly adsorbed on lattice sites, a Beer's law attenuation of the specular beam occurs with increasing adsorbate coverage. A cross section for diffuse scattering can be assigned, based on the attenuation of specular intensity and a knowledge of adsorbate coverage [Eq. (4.1)].

$$I/I_0 = (1 - \theta)\Sigma_A n_s \qquad\qquad (4.1)$$

where θ is the adsorbate coverage, n_s is the number of substrate lattice sites, and Σ_A is the adsorbate cross section. This cross section is typically quite large for molecular adsorbates on atomically smooth surfaces, ranging up to 150 Å^2 for CO on Pt(111) (greater than the gas phase He–CO scattering cross section).

For completely random adsorption, with no attractive or repulsive interactions between adsorbates, a plot of the logarithm of the normalized specular reflectivity versus coverage is a straight line. Positive or negative deviations from this behavior (Fig. 4.2) indicate attractive or repulsive interactions between the adsorbate species. The detailed behavior of the reflectivity as a function of exposure and surface temperature can provide a great deal of information about the dynamics of the adsorption process. Common situations are summarized in Figure 4.2. Extensive discussion of the method is provided in the monograph by Poelsema and Comsa, and in a number of papers in the literature.[2] The examples discussed in this chapter illustrate the sort of information about adsorption dynamics available using this powerful method.

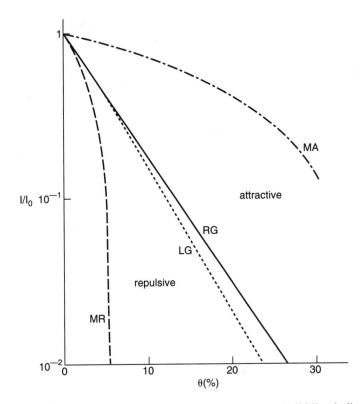

Figure 4.2. He specular intensity versus adsorbate coverage. Solid line indicates *random 2D gas* (RG) adsorption. MR indicates maximum repulsion, MA indicates maximum attraction, and LG is that predicted for a lattice gas model of adsorption. After ref. 1.

4.2 Illustrative Examples

The dynamics of three adsorption systems of varying complexity will be discussed in this chapter. The first is the adsorption of a rare gas adlayer, specifically Xe, on the Pt(111) surface. The coadsorption of CO and H_2 on the Pt(111) and the more open Fe(111) surface will then be discussed, with emphasis on adsorbate–adsorbate interactions in this system. Finally, the example of homoepitaxial growth [Pt on Pt(111)] as an example of adsorption dynamics will be considered in detail.

4.2.1 Xe Adsorption on Pt(111)

Weakly adsorbed, or physisorbed, systems can be very effectively investigated using the TEAS method. An excellent example of the use of TEAS to monitor the adsorption process in a weakly bound system is the adsorption of Xe on the nearly defect free Pt(111) surface.[3] The well-characterized Pt(111) surface is held at a low temperature, and the specular helium reflectivity is monitored while the surface is exposed to a low-pressure background of Xe gas. The measured specular helium intensity decreases rapidly with Xe exposure, and displays a very characteristic change in slope at a relatively low exposure. For increasing exposure, the measured intensity continues to decrease linearly with this new, smaller slope, over a wide exposure range. For the highest exposures, some deviation from this linear behavior is observed. This behavior suggests that initial adsorption is into a randomly adsorbed, mobile phase. As coverage increases, condensation (island formation) occurs, at a critical coverage characteristic of the surface temperature. The extensive linear portion of the exposure curve indicates that Xe atoms adsorb and are directly incorporated into the growing island, which may become quite large. The upward deviation from linearity at still higher exposures indicates that the Xe adatoms display an attractive interaction on this substrate.[4]

A careful examination of the initial stages of this adsorption process at a series of different surface temperatures provides more information about the phase transition corresponding to island formation. The initial part of the Xe adsorption curve for various temperatures is shown in Figure 4.3. The change in slope at increasing coverages for increasing surface temperatures and the fact that the curves are parallel for coverages above the critical coverage indicate the formation of two-dimensional (2D) islands of Xe on the Pt(111) surface. A $(\sqrt{3} X \sqrt{3})R30°$ LEED pattern is observed in the coverage range above this critical coverage, suggesting the formation of a commensurate 2D solid. At saturation coverage, the hexagonal incommensurate compression structure is observed. The slope of the initial adsorption curve, at coverages below the critical value, can be used to derive the cross section for He scattering by the randomly adsorbed Xe atoms in the lattice gas. This value is found to be 120

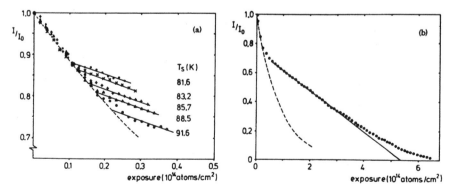

Figure 4.3. Relative specular He intensity versus Xe exposure. (a) Low coverage regime for various T_s. (b) Complete curve at $T_s = 90.1$ K. After ref. 3.

Å^2. This large cross section makes TEAS an especially appropriate tool for monitoring phase transitions in these weakly adsorbed systems, where 2D condensation occurs at extremely low coverages.

The shape of the adsorption curves is independent of large variations in the Xe pressure used for the exposure. Thus the 2D lattice gas and the 2D solid appear to be in dynamic equilibrium, and standard thermodynamic treatments can be used to derive thermodynamic properties of the system. For example, the 2D heat of vaporization is determined to be 1.1 kcal/mol. Xe on Pt(111) isobars can be measured by monitoring the specular He intensity as a function of temperature for a range of background Xe pressures.[4] Data of this type are shown in Figure 4.4. Using the Clausius–Clapeyron equation, these data can be analyzed to extract the isosteric heat of adsorption of Xe on the Pt(111) surface as a function of Xe coverage. This heat of adsorption is found to be 6.4 kcal/mol at zero coverage, and to increase with coverage to about 6.7 kcal/mol at a Xe coverage of 5%. A phase diagram for the Xe/Pt(111) system can also be derived using these data.

At higher coverages, He atom diffraction methods can be used to further explore the phase transition dynamics of this very interesting system.[5] The intensity of specific diffraction features corresponding to the commensurate, incommensurate, and rotated Xe overlayer structures is used to monitor the presence and concentration of each of the three structures. Again, the equilibrium nature of the system, over the coverage, temperature, and background Xe pressure

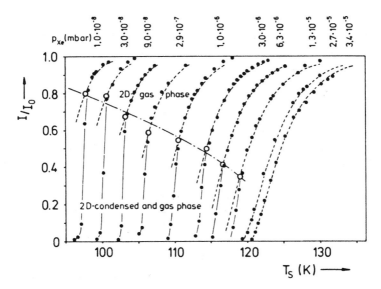

Figure 4.4. Xe on Pt(111) isobars, plotted as I/I_0 versus T_s. Measured data points are denoted by solid circles. The dashed curves show calculated isobars representing the Xe 2D gas phase. The open circles denote the Xe 2D gas-condensed phase transition. After ref. 4.

ranges explored, allows the extraction of thermodynamically meaningful quantities. Direct evidence for the incommensurate to commensurate phase transition is obtained, with a critical transition temperature of 58 K observed for Xe coverages around 0.33. Helium atom scattering provides a highly detailed picture of the adsorption and the adsorbate dynamics of this simple physisorption system. A similar approach is useful in the study of the more complex and chemically interesting examples presented in the following sections.

4.2.2 CO and H_2 Coadsorption on Pt(111)

The adsorption of CO and H_2 individually on the Pt(111) surface has been well studied. Carbon monoxide adsorbs molecularly on the Pt(111) surface, with a sticking coefficient near unity at low coverages.[6] The heat of adsorption is about 30 kcal/mol at low coverage, and decreases to about 20 kcal/mol at saturation. Normal saturation coverage corresponds to a half monolayer coverage, and a c(4 $\times$ 2) LEED pattern is exhibited by the ordered overlayer. Exposure of the saturated surface to relatively high CO pressures has been observed to result in a "dense phase" of CO on the surface with a higher order LEED pattern and a coverage of 0.65 monolayer. The associatively adsorbed molecules exhibit repulsive molecule–molecule interactions over the entire coverage range, and no evidence of CO island formation is obtained.[7] HREELS studies show a sharp single loss peak at 2200 cm^{-1}, and in combination with the LEED data suggest that the CO molecules are randomly adsorbed into equivalent 3-fold hollow sites on the Pt(111) surface as coverage increases.[8]

He scattering studies of CO adsorption on the Pt(111) surface indicate repulsive interactions between the molecules even at very low coverages. The CO molecules behave as perfectly diffuse scatterers for the incident He beam, with an effective cross section on the Pt(111) surface of 150 Å^2. The specular He reflectivity of the CO exposed Pt(111) surface rapidly falls below the detection limit, for coverages greater than about 20% of saturation.[7] (see Fig. 4.5).

Hydrogen adsorption on the Pt(111) surface is somewhat more complex. Molecular hydrogen readily dissociates on the Pt(111) surface, resulting in atomic hydrogen coverage of the surface.[9] Detailed studies suggest that dissociation occurs at defect sites, and that the perfect surface is relatively inactive for H_2 dissociation.[10] Even the most carefully prepared surfaces are not defect free, however, and the Pt(111) surface eventually becomes covered with atomic hydrogen when exposed to H_2. No ordered LEED patterns are observed for the hydrogen adsorption system. Thermal desorption shows a broad second order desorption peak at 350 K, and provides an estimate of 55 kcal/mol for the enthalpy of adsorption of hydrogen on the Pt(111) surface.[11]

TEAS studies of hydrogen adsorption on this surface indicate a rapid decrease in specular reflectivity with increasing hydrogen exposure, as is seen for CO adsorption on this surface. The effective cross section is considerably smaller, however, and the reflectivity does not go rapidly to zero.[12] Rather, it passes

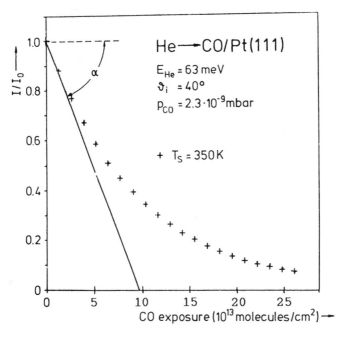

Figure 4.5. He specular intensity versus CO exposure. After ref. 1.

through a minimum with increasing coverage and increases to a saturation reflectivity characteristic of the atomic hydrogen overlayer. This value is about 60% of the clean surface reflectivity. This behavior is illustrated in Figure 4.6, and can be understood in terms of a simple model of scattered He beam interference from the clean and adsorbate covered regions of the surface. This model is illustrated in Figure 4.7, and suggests a height difference between the covered and uncovered regions of the surface of 0.18 Å, and an effective H adatom (or vacancy) scattering cross section of about 25 Å^2. The dissociative sticking coefficient for hydrogen on Pt(111) decreases markedly with coverage, requiring exposures of 10^4 to 10^5 Langmuirs to achieve saturation coverage on this surface.

The contrasting behavior of CO and H$_2$ adsorbed on Pt(111) as probed by TEAS provides an opportunity to probe the nature of coadsorption in this system.[13] Figure 4.8 shows the specular He reflectivity for a surface that is preexposed to CO and subsequently saturated with hydrogen. Initial exposure to CO results in a sharp drop in reflectivity to a value of about 25% of the clean surface reflectivity. This corresponds to a random CO coverage on the Pt(111) surface of about 0.14 monolayer. This surface is then exposed to hydrogen, and the reflectivity drops through a minimum and begins to increase again as observed for hydrogen adsorption on the clean surface. In this case, however, the reflectivity at saturation *is greater than* the reflectivity of the initial CO exposed surface. This suggests that the adsorbed hydrogen is forcing the randomly pread-

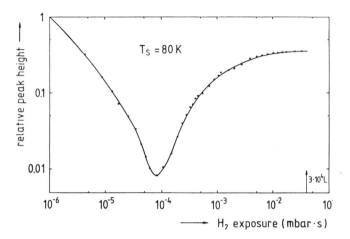

Figure 4.6. He specular intensity versus H_2 exposure. After ref. 1.

Method

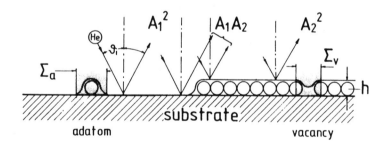

He specular peak height:

$$I/I_0 = A_1^2 + A_2^2 + 2f\cos\varphi\, A_1 A_2$$

$$\varphi = 2\pi \cdot \frac{2h\cdot\cos\vartheta_i}{\lambda}$$

Figure 4.7. Model for specular He interference between bare and H covered Pt(111). After ref. 1.

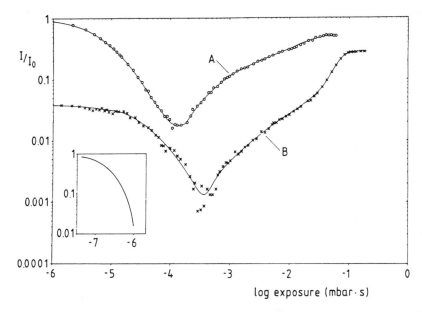

Figure 4.8. Specular He intensity (I/I_0) from Pt(111) as a function of exposure to H_2 for an initially clean (A) and a CO precovered (B) surface. Inset shows I/I_0 versus CO exposure for the initially clean Pt(111) surface. After ref. 13.

sorbed CO molecules to form islands of higher than normal saturation coverage. The effective cross section per adsorbed CO is reduced because of the hydrogen saturation induced overlap of the CO scattering cross sections. A simple geometric model can be invoked to predict the expected hydrogen saturation reflectivity for varying CO precoverages. The results of this simple model prediction are shown in Figure 4.9. In addition, LEED evidence for the "dense phase" CO overlayer is obtained for CO coverages well below 0.6 monolayer, but that have high local densities due to this hydrogen saturation-induced islanding. Thus, clear evidence is obtained for the formation of adsorbate islands in a system where individual adsorbate–adsorbate interactions are purely repulsive. This observation suggests that H–CO adsorbate–adsorbate repulsive interaction energies are greater than either CO–CO or H–H interaction energies on the Pt(111) surface.[13] Further TEAS studies provided information about island size and size distribution in this system, and about the mobility of adsorbed CO in the presence of coadsorbed hydrogen.[14]

4.2.3 CO and H_2 Coadsorption on Fe(111)

The Fe(111) surface is much more open and atomically rougher than the Pt(111) surface discussed in the previous section, and distinct differences are observed in the adsorption of CO and H_2 on this surface. Dissociative adsorption of CO

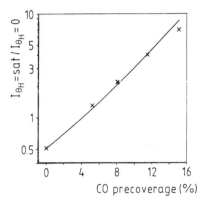

Figure 4.9. Specular He intensity at hydrogen saturation normalized to specular intensity following CO preexposure as a function of CO precoverage. Solid line calculated based on geometric model described in text. After ref. 13.

on the Fe(111) surface is significant, and occurs readily for temperatures above ambient.[15] Carbon monoxide molecularly adsorbed at lower temperatures both dissociates and desorbs molecularly as the surface is warmed. Dissociatively adsorbed CO recombines and desorbs at still higher temperatures. HREELS and thermal desorption studies suggest that molecular CO is adsorbed in three distinct sites on the open Fe(111) surface,[16] with three distinct CO stretching frequencies observed in the vibrational spectrum.

Perhaps because of these distinct adsorption sites, and certainly because of

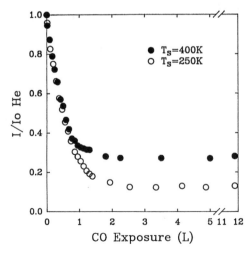

Figure 4.10. Relative scattered He intensity versus CO exposure on Fe(111) where θ_i = 40°. Surface temperatures are as given in the legend. After ref. 17.

the atomic roughness of the Fe(111) surface, CO adsorption probed by TEAS is less readily interpreted than in the Pt(111) case.[17] Figure 4.10 shows the normalized specular He intensity as a function of CO coverage for adsorption on the Fe(111) surface at 250 and 400 K. The intensity drops with coverage but does not rapidly become undetectable. Rather, the specular intensity levels off at about 20% of the initial value at 250 K, and somewhat higher at 400 K. Clearly, CO adsorbed on the Fe(111) surface exhibits different He scattering behavior than CO on Pt(111), and the effective scattering cross section (56 Å^2) is much less than the cross section observed for CO on Pt(111). This data can be fit to a diffuse scattering model resulting in the effective cross section mentioned above.

Hydrogen adsorption on the Fe(111) surface has been less well studied. Thermal desorption studies indicate three distinct dissociative adsorption sites,[18] which may be analogous to the sites identified for CO adsorption, although there is no conclusive evidence for this. No vibrational spectroscopic measurements or detailed structural determinations have been carried out. The most strongly bound adstate fills first with increasing exposure, and the second and third states are progressively less strongly bound. As with the Pt(111) surface, very high exposures are necessary for complete saturation of the surface. There remains some uncertainty about the absolute coverage values, but consistent arguments can be made to support a saturation coverage of 1.378×10^{15} hydrogen atoms per cm^2.[19]

This added complexity is reflected in the TEAS measurements of this system. Figure 4.11 illustrates the surface reflectivity as a function of hydrogen coverage

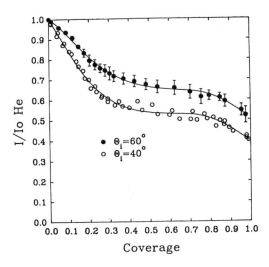

Figure 4.11. Relative scattered He intensity as a function of H coverage for two incident angles at a surface temperature of 173 K. Incident angles are as given in the legend. The solid lines are the fit of the model described in ref. 20.

for adsorption at 173 K. The observed behavior cannot be explained by simple perfect diffuse or nonperfect diffuse scattering models. Rather, the distinct ad-sites must be ascribed distinct effective scattering cross sections. In combination with assumptions about the thermal distribution of adatoms among these sites as a function of increasing coverage, a reasonable fit to the TEAS observations is obtained.[20]

Coadsorption of CO and H_2 on the Fe(111) surface is an interesting system technologically as well as from the viewpoint of basic understanding. Fischer–Tropsch chemistry on iron-based catalysts takes place by way of the surface interaction of CO and hydrogen. In spite of the technological relevance little work has been carried out on this coadsorption model system. Studies have indicated that hydrogen will not adsorb on the CO predosed surface, and that CO will readily displace adsorbed hydrogen. No evidence has been obtained for reaction of CO with hydrogen at the low coverages and temperatures typical of these model studies, while reaction certainly takes place under operating conditions typical of Fischer–Tropsch synthesis.

TEAS provides direct information about the subtle interactions involved in this coadsorption system. Figure 4.12 illustrates the specular He intensity as a function of hydrogen exposure for surfaces preexposed to varying quantities of CO. The observed drop in intensity clearly indicates that hydrogen will adsorb on the CO predosed surface. The reverse adsorption order, H_2 preexposure followed by CO saturation results in TEAS data similar to CO adsorption on the

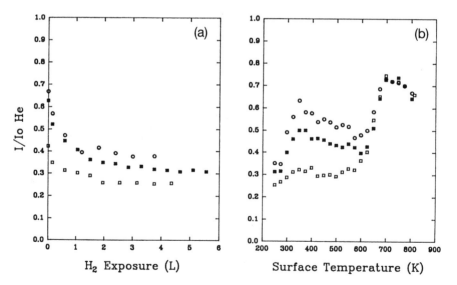

Figure 4.12. (A) Specular He intensity versus H_2 exposure for CO predosed Fe(111) surfaces at $T_s = 250$ K. Nominal CO preexposures are (○) 0.1 L, (■) 0.25 L, (□) 0.5 L. (B) Specular He intensity versus surface temperature following hydrogen exposure of surfaces in (A). (B) is uncorrected for Debye–Waller effects. After ref. 21.

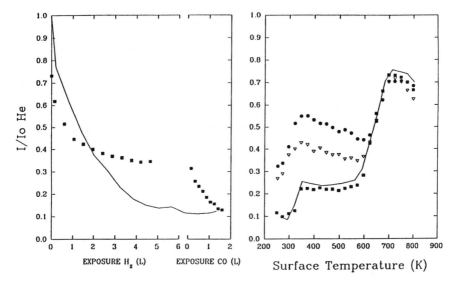

Figure 4.13. Specular He intensity from Fe(111) exposed to 0.1 L CO, followed by H_2 exposure until constant, followed by various CO exposures. Final CO exposures are (●) 0.05 L, (▽) 0.1 L, (■) 2 L. Solid lines in (A) and (B) are for CO adsorbed on the clean Fe(111) surface. (B) is uncorrected for Debye–Waller effects. After ref. 21.

clean surface, suggesting that CO does, in fact, displace hydrogen from the surface. When the CO preexposed, hydrogen-saturated surface is warmed and the reflectivity monitored, interesting behavior is observed. The specular intensity increases significantly with the onset of hydrogen desorption, but then relaxes rapidly and to an extent proportional to the CO precoverage, prior to eventual CO desorption and reflectivity recovery to the clean surface value (see Fig. 4.13). This relaxation of the reflected intensity suggests a spreading of the adsorbed CO from a densely covered region to a more dilute coverage, effectively increasing the per molecule scattering cross section. While the effect is not as dramatic or as theoretically describable as that observed on the Pt(111) surface, there is clear evidence for hydrogen–carbon monoxide repulsive interactions on this surface as well. Although this interaction is apparently not large enough to cause adsorbate segregation, it does restrict the CO mobility and inhibit CO dissociation.[21]

4.2.4 Pt on Pt(111) Epitaxial Growth

As a final example of the detailed study of adsorption dynamics consider the epitaxial growth of platinum on a well-characterized Pt(111) substrate. This provides an understandable model system for the important processes of homo- and heteroepitaxial thin film growth. The phenomonology and morphology of thin film growth are longstanding research topics of clear technological importance.

The subject has been addressed by electron diffraction, field ion microscopy, and electron spectroscopy among other methods. The present discussion will again focus on the use of TEAS to probe the adsorption and epitaxial growth process.

Early work by Ibáñez and co-workers[22] indicated that TEAS could be used to monitor film growth on metal substrates. In this case, the growth of Pb layers on a copper substrate was studied, and evidence for layer-by-layer growth was obtained from the observed specular He intensity oscillations. More recent work by the Jülich group has concentrated on the simpler homoepitaxial system Pt on Pt(111)[23] and uncovered a wealth of detail about the dynamics of the adsorption and growth process.

This is illustrated by the series of TEAS measurements for different substrate temperatures and fixed platinum deposition rate shown in Figure 4.14. At high temperatures, the specular reflectivity stays essentially constant with platinum exposure, indicative of step propagation growth. The adsorbed adatoms are

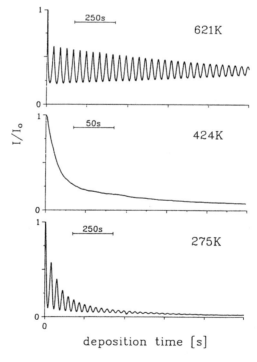

Figure 4.14. The normalized He specular peak height vs Pt on Pt(111) deposition time measured at three different surface temperatures. All other parameters including the deposition rate (confined between 1/40.6 and 1/36 ML/sec) are identical. The oscillatory behavior at high and low temperature is evidence for layer-by-layer growth (denoted in the text by $2D_h$ and $2D_l$ growth, respectively); the monotonic decrease in the intermediate region indicates 3D growth. After ref. 23.

highly mobile and are incorporated into step edges rapidly on the time scale of the measurement. Smooth growth morphology results.

When the substrate temperature is lowered, oscillations in the specular reflectivity are observed. These oscillations correspond to a layer-by-layer growth regime, and result from the interference between the He scattered from the substrate and the partially completed overlayer, analogous to the previously discussed nonperfect diffuse scatterer model for hydrogen adsorption on Pt(111). In the case of Pt on Pt(111), however, the completed monolayer has the same reflectivity as the initial substrate, so oscillations in reflected intensity are observed as layer after layer is deposited. This growth mode relies on sufficient adatom and vacancy mobility so that adatoms deposited in the third layer move to the edge of the underlying island and are incorporated in the second layer rather than nucleating growth of a third layer.

As substrate temperature is reduced still further, adatom mobility drops and nucleation of third, fourth, and higher layers takes place. This approach to three-dimensional growth is indicated by the loss of oscillations and the rapid drop in specular intensity with increasing exposure. The resulting growth surface exposes several different monolayer levels, and is microscopically quite rough. The drop in specular intensity is due to a monotonic increase in defect density as more platinum is deposited, specifically an increase in the step edge length of the increasing number of exposed layers. This growth mode is often termed "Poisson distributed" or 3D growth, and up to seven individual layers may be exposed simultaneously.

As the substrate temperature drops below room temperature, oscillations in the TEAS specular intensity are again observed. This "reentrant" layer-by-layer growth mode is unexpected based on the simple mobility arguments discussed above for the high temperature oscillations, and led to the interesting proposal of Egelhoff and Jacob[24] that adatom mobility at these low substrate temperatures must be due to the latent heat of condensation of the deposited atom. In the present study of homoepitaxial growth of platinum on platinum, deposited adatoms are mobile over the entire temperature range of the study, and the deposition rate is high enough that supersaturation conditions exist, so island formation will occur. When adatoms are deposited randomly on the surface, island nucleation occurs when the surface adatom pressure is high enough. Once formed, these islands grow rapidly when adatom mobility is high, and island coalescence occurs before a new island can be nucleated on top of the growing one. This is the case for ideal 2D growth, the $2D_h$ region of this study. If new islands nucleate on top of the underlying island before coalescence, due to a barrier to adatom mobility over the edge of the growing island or to a build up of a critical nucleation pressure on the top of the growing island, then 3D growth occurs. The reentrant $2D_l$ growth regime observed when the temperature is further decreased is related to reduced island size or to a change in island morphology because of lower adatom mobility in this temperature regime. Small islands of irregular shape are proposed to keep the critical adatom density low on the top of the growing islands, preventing the formation of supercritical nuclei and quasi 2D growth in the lower temperature range. This model suggests clear differences in the shape and size distribution of the growing islands in the three different growth modes.

This suggestion was elegantly tested by further work from the Jülich group. Examination of the detailed shapes of the TEAS oscillations, concentrating on the in-phase data, showed a temperature-dependent asymmetry in the shapes of the oscillations.[25] In the $2D_h$ regime, both in-phase and anti-phase oscillatory behavior is very long lived and of essentially constant intensity. As many as 150 oscillations can be seen, with eventual damping due to spatial inhomogeneities in the deposition source rather than actual roughening of the surface by a change in growth mode. In this high temperature range, neither the in-phase nor the antiphase intensity recovers to the initial specular reflectivity, suggesting that the ideal 2D growth behavior with only two exposed layers is not attained, but that the growth front likely involves about four exposed layers. In the lower temperature 3D growth regime, the antiphase data drop exponentially as is expected for Poisson distributed growth, with seven or eight layers exposed simultaneously. As the substrate temperature is lowered further, and reentrant $2D_1$ growth is observed, both the in-phase and antiphase data show characteristic oscillations that are rapidly damped after only 15 or 20 cycles. The antiphase data in both 2D growth regimes are essentially symmetric with deposition time, nearly parabolic in shape as would be expected for the quadratic dependence of scattered intensity on coverage of the ideal 2D growth mode. The in-phase data appear highly asymmetric, however, where the intensity minimum is reached well before the coverage corresponding to half a monolayer deposited. This asymmetry depends as well on the deposition temperature. In the low temperature $2D_1$ region, the asymmetry is small and nearly independent of temperature. In the higher temperature 2D region, the asymmetry is significant and grows rapidly with increasing surface temperature. This asymmetry suggests that in this temperature range, island coalescence does not coincide with the minimum in the in-phase scattered intensity. Large islands must be growing at the expense of smaller ones, which disappear as growth continues. This is a description of Ostwald ripening, and is likely to occur at these higher growth temperatures, where adatoms can detach themselves from the edge of a small island and move across the substrate to be incorporated in the growing larger island. In the lower temperature reentrant 2D growth regime, island morphology is such that smaller, rougher islands are present and island coalescence corresponds to the half monolayer point.

A consistent and beautiful confirmation of the homoepitaxial growth processes inferred from the TEAS measurements has also been provided by Comsa and co-workers.[26] In a complementary series of STM studies, the three growth modes of platinum on platinum have been confirmed, and further insights into the growth morphology have been obtained. Scanning tunneling microscopy has been used to directly image the surfaces that result following deposition of platinum on Pt(111) under conditions corresponding to the $2D_h$, 3D, and $2D_1$ growth modes. These results are summarized in the STM topographs of Figure 4.15.

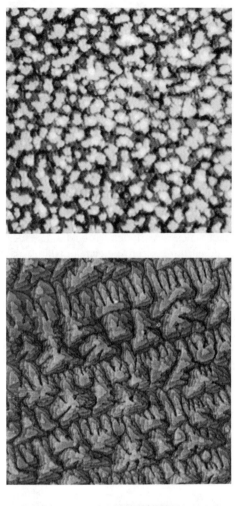

Figure 4.15. STM topograph 1000 × 1000 Å and 4000 × 4000 Å (b,c); 2.4 ± 0.3 ML platinum has been deposited on a Pt(111) sample, the sample temperature was $T = 205$ K (a), $T = 425$ K (b), and $T = 628$ K (c). After ref. 26.

Figure 4.15a shows the morphology that results following the deposition of 2.4 ML of platinum on the Pt(111) surface at 205 K. Figure 4.15b and c shows the morphology following deposition of the same amount of platinum at substrate temperatures of 425 K and 628 K, respectively. The difference in morphology in these three cases is striking. In the $2D_h$ regime (Fig. 4.15c), large convex islands are present. About four layers can be clearly seen, two of them dominating the imaged area. Small nuclei on the top-most layer are visible, as are small regions of incomplete lower-most layer. As coverage increases, it is clear that coalescence of these large, rounded islands occurs before half monolayer coverage is reached, and that these islands have second and even sometimes third layer islands nucleated on top of them. Examining the morphology of the $2D_l$ growth regime (Fig. 4.15a), it is seen that at the same coverage, the situation is much different. Two-dimensional islands are again observed, but they are much smaller, and much rougher in aspect. These small, dendritic island shapes are consistent with the arguments proposed earlier concerning critical island size and mobility of adatoms over the edge of the growing island. Clearly here coalescence will occur very nearly at the half monolayer point, and nucleation of second and third layer islands on these small first layer islands is not very likely. The 3D growth regime (Fig. 4.15b) displays a markedly different morphology than that seen for the other two growth modes. Here the surface is covered after the deposition of 2.4 ML of platinum with easily recognizable pyramids, exposing up to seven or eight layers of platinum. This is completely consistent with the Poisson distributed 3D growth expected based on the TEAS results discussed above. The amazing results of this investigation demonstrate the power of combining the quantitative aspects of diffraction probing of surface dynamics with the qualitative visual aspects of the real space imaging of surface processes provided by STM.

4.3 Comments

In this chapter we have seen the power and range of detail that are available from the use of TEAS to probe the dynamics of adsorption and film growth in heterogeneous systems. The extreme sensitivity of this method combined with the reasonably straightforward interpretation possible for the specular scattering intensity make this approach one that will be increasingly used to investigate heterogeneous reaction dynamics. As an in situ probe of surface coverage it has unrivaled sensitivity, and can be applied to situations in real time, as has been seen for the homoepitaxial growth example explored in detail here. It is likely that this method will be applied more regularly to more complex systems, such as the coadsorption systems described above. In addition, TEAS is a useful in situ probe of surface coverage that can be effectively coupled with gas phase detection of desorbed product in modulated molecular beam studies of heterogeneous reaction processes.[27] It is also likely that TEAS will be applied more regularly to interactions on more open surfaces such as the Fe(111) work used as an example in this chapter. The method is not restricted to highly reflective substrates, and its more widespread application to more complex and to more open substrates should be expected.

Clearly scanning tunneling microscopy and the other scanning probe microscopies will be applied more regularly to the study of the dynamics of adsorption, epitaxial growth, and adsorbate interactions. The sort of understanding that is available from the direct imaging of real space surface structures is difficult to duplicate with any other method. As the method is developed experimentally, its very widespread use to examine more and more complex systems will occur. Recent work has investigated the spatial dependence of surface reaction dynamics, probing the effect of island and overlayer structures on the kinetics and dynamics of simple surface reactions.[28] Such measurements require the careful experimental control and variation of surface temperature while scanning. Developments in the theoretical understanding of the tunneling mechanism will also be needed to understand imaging of molecular adlayers, and to interpret changes in the images that convolute topography with changes in surface electronic structure. This is a very active area of research, however, and such theoretical and instrumental developments are well underway. Scanning probe microscopy will soon be one of the principal tools for the investigation of the detailed dynamics of heterogeneous systems.

References

1. B. Poelsema and G. Comsa, Scattering of thermal energy atoms from disordered surfaces. *Springer Tracts in Modern Physics 115,* Springer, Berlin, 1989.

2. B. Poelsema, S. T. de Zwart, and G. Comsa, *Phys. Rev. Lett.* **51,** 522 (1983).

   A. M. Lahee, J. R. Manson, J. P. Toennies, and Ch. Wöll, *J. Chem. Phys.* **86,** 7194 (1987).
   H. Xu, Y. Tang, and T. Engel, *Surface Sci.* **255,** 73 (1991).

3. B. Poelsema, L. K. Verheij, and G. Comsa, *Phys. Rev. Lett.* **51,** 2410 (1983).

4. B. Poelsema, L. K. Verheij, and G. Comsa, *Surface Sci.* **152/153,** 851 (1985).

5. K. Kern, R. David, R. L. Palmer, and G. Comsa, *Phys. Rev. Lett.* **56,** 620 (1986).

6. H. Steininger, S. Lehwald, and H. Ibach, *Surface Sci.* **123,** 264 (1982).

7. B. Poelsema, R. L. Palmer, and G. Comsa, *Surface Sci.* **136,** 1 (1984).

8. B. E. Hayden, K. Krezschmar, A. M. Bradshaw, and R. G. Greenler, *Surface Sci.* **149,** 399 (1985).

9. K. Christmann and G. Ertl, *Surface Sci.* **60,** 365 (1976).

10. S. L. Bernasek and G. A. Somorjai, *J. Chem. Phys.* **62,** 3149 (1975).

11. K. E. Lu and R. R. Rye, *Surface Sci.* **45,** 677 (1974).

12. B. Poelsema, L. Brown, K. Lenz, L. K. Verheij, and G. Comsa, *Surface Sci.* **171,** L395 (1985).

13. S. L. Bernasek, K. Lenz, B. Poelsema, and G. Comsa, *Surface Sci.* **183,** L319 (1987).

14. K. Lenz, B. Poelsema, S. L. Bernasek, and G. Comsa, *Surface Sci.* **189/190,** 431 (1987).

15. C. E. Bartosch, L. J. Whitman, and W. Ho, *J. Chem. Phys.* **85,** 1052 (1986).

16. U. Seip, M. C. Tsai, K. Christman, J. Küppers, and G. Ertl, *Surface Sci.* **139**, 29 (1984).

17. P. Jiang, M. Zappone, and S. L. Bernasek, *J. Chem. Phys.* **99**, 8120 (1993).

18. F. Boszo, G. Ertl, M. Grunze, and M. Weiss, *Appl. Surface Sci.* **1**, 103 (1977).

19. R. W. Pasco and P. J. Ficalora, *Surface Sci.* **134**, 476 (1983).

20. P. Jiang, M. Zappone, and S. L. Bernasek, *J. Chem. Phys.* **99**, 8126 (1993).

21. S. L. Bernasek, M. Zappone, and P. Jiang, *Surface Sci.* **272**, 53 (1992).

22. L. J. Gómez, S. Bourgeal, J. Ibáñez, and M. Salmerón, *Phys. Rev. B* **31**, 2551 (1985).

23. R. Kunkel, B. Poelsema, L. K. Verheij, and G. Comsa, *Phys. Rev. Lett.* **65**, 733 (1990).

24. W. F. Egelhoff, Jr. and I. Jacob, *Phys. Rev. Lett.* **62**, 921 (1989).

25. B. Poelsema, A. F. Becker, G. Rosenfeld, R. Kunkel, N. Nagel, L. K. Verheij, and G. Comsa, *Surface Sci.* **272**, 269 (1992).

26. M. Bott, T. Michely, and G. Comsa, *Surface Sci.* **272**, 161 (1992).

27. K. A. Peterlinz, T. J. Curtiss, and S. J. Sibener, *J. Chem. Phys.* **95**, 6972 (1991).

28. T. A. Land, T. Michely, R. J. Behm, J. C. Hemminger, and G. Comsa, *J. Chem. Phys.* **97**, 6774 (1992).

Surface Diffusion

Diffusion is an essential elementary step in surface reaction processes. A microscopic understanding of diffusion on surfaces is clearly essential to an overall understanding of heterogeneous reaction dynamics. Once a reactant molecule is trapped on the surface, it often must migrate to a reactive site or a particular defect structure on the surface before further reaction can take place. Or the adsorbed reactant must diffuse across the surface to encounter another reactant species with reaction as the result. Self-diffusion of adatoms or clusters on a solid surface is an important step in crystal growth from the vapor, or in epitaxial growth of metals on metal or semiconductor substrates. The surface formed product molecule may also diffuse over the surface to reach a defect, active site, or region of the surface from which desorption can take place. A complete description of heterogeneous reaction dynamics requires a careful microscopic understanding of surface diffusion, including mechanisms and diffusion rates for important representative systems.

Unfortunately, the accurate determination of microscopic diffusion parameters on characterized surfaces is experimentally very difficult. Over the years a number of approaches have been developed, ranging from direct observation of metal on metal diffusion using field ion microscopy[1] to the very elegant emission current fluctuation measurements pioneered by Gomer,[2] to grating relaxation based measurements applied by Bonzel and co-workers.[3] Two particular approaches will be discussed in detail in this chapter. Additional approaches, and an extensive tabulation of adsorbate and self-diffusion data, are discussed in a recent review by Bonzel.[4]

5.1 Experimental Methods

The two case studies discussed in this chapter rely on the use of surface perturbation methods to measure diffusion parameters. In the first example laser irradiation of a well-characterized, adsorbate covered surface is used to create an
adsorbate free region of defined geometry on the surface. Diffusional refilling
of this bare region from the surrounding adsorbate covered region of the surface
is then monitored by subsequent laser desorption to measure the amount of
adsorbate present in the initially irradiated area. Repetition of this sequence with
varying time delays between the initial adsorbate removal and subsequent probe
pulses provides a direct measure of the adsorbate diffusion rate on a surface

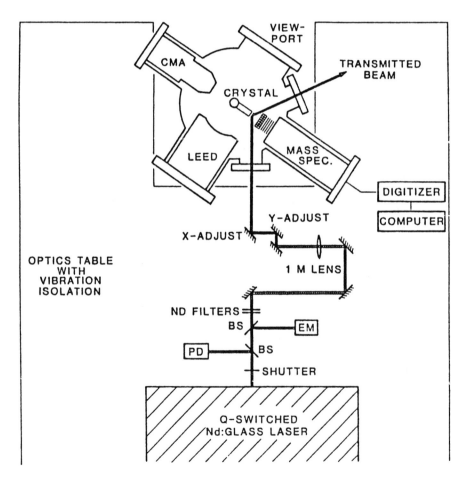

Figure 5.1.   Schematic diagram of laser desorption apparatus for measuring surface
diffusion. After ref. 9.

which can be carefully prepared and characterized.[5] Figure 5.1 illustrates this experimental approach schematically.

This experimental approach has a number of advantages for measuring adsorbate diffusion on surfaces. The first, and most important, is its widespread applicability. Diffusion of most chemically interesting adsorbates on most metallic, and many semiconductor surfaces can be probed. Adsorbate–substrate combinations are limited to systems in which laser-induced thermal desorption can be used to nondestructively desorb the adsorbate species. Systems exhibiting reasonable chemisorption energies (20–70 kcal/mol) are amenable to this sort of probing. The surface itself can be carefully prepared and characterized, using the entire arsenal of UHV spectroscopic probes. Substrate composition is not limited to refractory metals as in field ion or field emission probes. The substrate temperature and adsorbate initial coverage can be varied and controlled over fairly wide ranges of conditions. The experiment itself is conceptually fairly simple, and the concentration versus time data obtained can be directly related to diffusion parameters using Fick's laws.

There are disadvantages to this approach as well, the principal one being the effect of the desorbing pulse on the substrate. Laser irradiation of metals, and the subsequent temperature transient, can introduce point defects and slip plane damage, even at fluences that are small compared to those sometimes used in laser desorption measurements.[6] The presence of defects in the irradiated area may affect subsequent diffusional refilling, and make it difficult to obtain the true diffusion rate for the initial defect free surface. Care must also be taken in the experimental design to minimize system volume and sample to detector distances to maximize sensitivity to the transient pressure rise generated by the probe pulse. An accurate knowledge of the geometry of the irradiated area must also be obtained, so that correct solutions to the diffusion equation describing the refilling process can be carried out. These disadvantages have been carefully considered by those who have developed this method, primarily George and coworkers,[7,9,10] resulting in diffusion measurements for a number of interesting adsorbate–substrate systems.

An alternative perturbation approach to the determination of adsorbate diffusion rates on well-characterized surfaces is the pulsed molecular beam method developed by Reutt-Robey et al.[8] In this case, as illustrated in Figure 5.2, a pulse of adsorbate, in a well-collimated molecular beam, is incident on a well-characterized substrate. The substrate is monitored in real time by a reflection IR spectrometer. Diffusion is then monitored by the change in intensity with time, of adsorbate vibrational frequencies characteristic of the adsorbate binding site. This approach was first elegantly applied to CO diffusion on platinum (111), where the CO stretching frequency for molecules adsorbed on the (111) terrace is some 30 cm^{-1} different than that for CO adsorption at a step edge. The results of this measurement are discussed in more detail below.

This approach shares some of the advantages of the laser desorption method described above. In particular, a wide range of adsorbate–substrate combinations is, in principle, amenable to study. The substrate itself can be carefully prepared

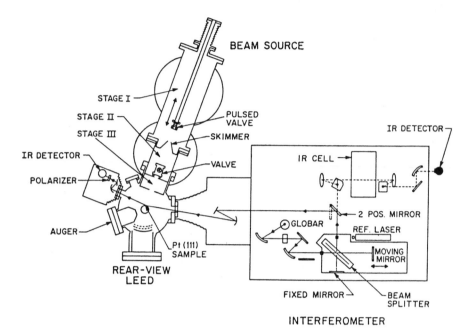

Figure 5.2. Schematic diagram of molecular beam infrared reflection absorption spectroscopy apparatus for measuring surface diffusion. After ref. 8.

and characterized. Initial coverages and surface temperature can be controlled. In the disadvantage column, the real choice of adsorbate–substrate combination depends on the existence of a system with IR distinguishable adsorption sites (terrace/step) and adsorbates that can be detected using reflection IR. The experimental apparatus, with pulsed molecular beam source and high quality reflection IR capability, is also not inexpensive.

5.2 Illustrative Examples

The two examples discussed in this chapter both deal with molecular diffusion on well-characterized transition metal surfaces of catalytic importance. Molecular diffusion on these metals is particularly difficult to study, and these examples illustrate the power of the two methods we concentrate on here. Other systems and other approaches have been used over the years, primarily providing information on metallic self-diffusion, or atomic adsorbate diffusion on refractory metals. A microscopic understanding of molecular diffusion on catalytically important metals is essential for a complete treatment of heterogeneous reaction dynamics. The diffusion of H_2 and CO on ruthenium is an important step in Fischer–Tropsch catalysis chemistry on this metal. Similarly the diffusion of CO on platinum is important for understanding the catalytic oxidation of CO, and provides another piece of detail in the microscopic description of this widely studied model heterogeneous reaction system.

5.2.1 Laser-Induced Desorption Measurement of Hydrogen and CO Diffusion on Ru(001)

The diffusion of dissociatively adsorbed hydrogen atoms on the Ru(001) surface has been probed using laser-induced thermal desorption. Initial experiments probed a hydrogen coverage that results from a 0.05 L H_2 exposure to the Ru(001) surface held at 150 K. Desorption measurements were then carried out over the temperature range from 260 to 330 K. Typical experimental diffusional refilling data obtained using this method is illustrated in Figure 5.3. For the experimental geometry used here (58° laser angle of incidence), an elliptical laser spot is created. Fick's second law is solved for this elliptical symmetry, and expected diffusional refilling curves were calculated. These are indicated as the solid lines of Figure 5.3, and assume a constant diffusion coefficient for this coverage and temperature range.

The experimental diffusion coefficient derived from these refilling curves can then be used to extract an activation energy for diffusion, and a preexponential factor assuming Arrhenius behavior for this temperature range. The activation barrier extracted in this manner is 4.0 ± 0.5 kcal/mol for hydrogen diffusion on the clean Ru(001) surface at low hydrogen coverage. The preexponential factor, D_0 derived from this treatment was found to be $D_0 = 6.3 \times 10^{-4}$ cm^2/sec. The adsorption energy of hydrogen on Ru(001) has been determined to be about 60 kcal/mol. Thus, the ratio of $E_{diff}/E_{ads} \cong 0.07$ on this surface, in relatively good agreement with this ratio measured for hydrogen on nickel[10] or tungsten.[11] The activation barrier, which is likely to represent the barrier for H atom motion between adjacent 3-fold sites via a 2-fold site on the Ru(001) surface, is also consistent with vibrational spectroscopy of H adsorbed in the threefold site,[12]

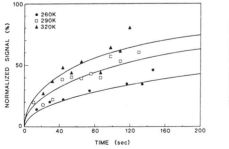
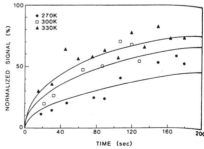

Figure 5.3. Diffusional refilling curves for hydrogen diffusion on Ru(001) for several surface temperatures. Solid lines are fits assuming a constant diffusion coefficient. After ref. 9.

and with conventional transition state theory for this diffusional motion.[13] Diffusional motion of H atoms on the Ru(001) surface likely occurs microscopically by random single hops between adjacent threefold sites on the surface.

George and co-workers have also examined the initial coverage dependence of hydrogen diffusion on this surface,[14] as well as differences in diffusion between hydrogen and deuterium.[15] Coverage-dependent measurements were carried out over the range of coverages from 15 to 85% of saturation. The measured hydrogen diffusion coefficient was found to be constant over this coverage range. This suggests that adsorbate–adsorbate interactions are negligible for this system. This coverage independence suggests that coverage-dependent features observed in thermal desorption and HREELS studies of this adsorption system[16] are likely to be due to the existence of two inequivalent hydrogen adsorption sites rather than significant repulsive lateral interactions among the adsorbate species.

Comparison of deuterium and hydrogen diffusion on the Ru(001) surface addresses the question of quantum mechanical tunneling as a possible mechanism for hydrogen diffusion on this surface. Measurements with deuterium were carried out as before, and diffusional refilling curves calculated based on a solution of Fick's second law.[15] The results of these measurements are shown in Figure 5.4. The measured activation barriers for diffusion were 3.6 ± 0.5 and 4.1 ± 0.5 kcal/mol for hydrogen and deuterium, derived from this Arrhenius

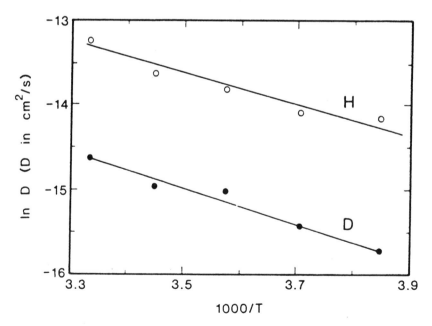

Figure 5.4. Arrhenius plots of diffusion coefficients for hydrogen and deuterium on the Ru(001) surface. After ref. 15.

plot. The clearly Arrhenius behavior observed for both isotopes suggests classical, rather than quantum-mechanical (tunneling) behavior. The ratio of diffusion preexponentials, D_0^H/D_0^D, is observed to be 1.5, very close to the classical prediction of 1.4 due to the $\sqrt{m}$ mass correction. Similarly, the small measured difference in activation energy for diffusion ($\sim$ 0.5 kcal/mol), after taking into account zero point energy differences for the two isotopes, suggests that tunneling accounts for a very small portion of the diffusion rate for hydrogen on the Ru(001) surface.

In contrast to the behavior of hydrogen on Ru(001), measurements of CO diffusion on this surface exhibit a strong coverage dependence.[17] Below a CO coverage of $\theta = 0.33$, the measured CO diffusion coefficient was found to be essentially independent of coverage. Between 33 and 60% of a monolayer initial CO coverage, the diffusion coefficient was strongly coverage dependent. Figure 5.5 summarizes the behavior of the measured CO diffusion coefficient as a function of initial CO coverage for two separate surface temperatures. The strong increase in diffusion coefficient in the high coverage regime is ascribed to strong repulsive CO–CO interactions at these coverages. A repulsive, pairwise CO–CO interaction energy of 1.4 kcal/mol was used to describe the coverage dependent diffusion behavior illustrated here. This is a significant fraction of the measured diffusion activation energy of 6–8 kcal/mol for this coverage range. This strong

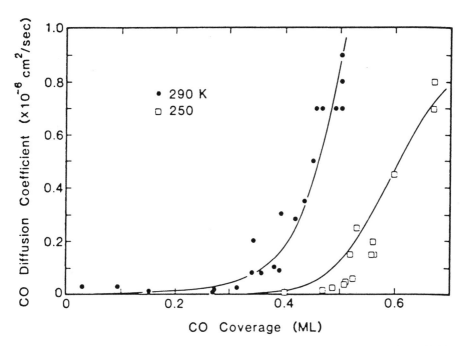

Figure 5.5. Surface diffusion coefficients for CO on Ru(001) as a function of CO coverage at 290 and 250 K. After ref. 17.

repulsive interaction is consistent with previous LEED, thermal desorption, and vibrational spectroscopic studies of this adsorption system.[18]

Laser-induced thermal desorption has proven to be an effective method for providing detailed microscopic diffusion information for important adsorbates on well-characterized single crystal substrates. In addition to the diffusion measurements on the clean Ru(001) surface described above, this approach has been used to investigate diffusion on surfaces modified by carbon, sulfur, and oxygen coadsorbates.[19] This sort of information is essential to developing an understanding of the detailed dynamics of surface reactions. This approach has also recently been applied to a series of n-paraffin molecules.[20,21] The method has also been applied to other substrates, including nickel,[22] rhodium,[23] and platinum[24] surfaces. As long as care is taken to avoid laser irradiation-induced defects, this approach is developing into a general and widely used method to measure adsorbate surface diffusion. A very clever modification of this approach has recently been reported.[25] In this case, an interference pattern from coincident laser pulses creates a desorption grating on an initially adsorbate covered surface. This spatial modulation in coverage is then monitored by second harmonic generated signal as the grating relaxes by diffusion of adsorbate. Carefully monitoring this signal as a function of time, surface temperature, and grating conditions can be used to derive microscopic diffusion parameters. This approach promises to have fairly general applicability.

5.2.2 Pulsed Molecular Beam Reflection IR Studies of CO Diffusion on Platinum

A so far less widely used approach, but an elegant one that also promises detail and generality in the study of surface diffusion, is the pulsed molecular beam-reflection IR method developed by Reutt-Robey et al.[8] In this approach, described briefly above, a pulsed beam of adsorbate, of 700 μsec duration, is incident on a clean stepped platinum surface. The time evolution of the vibrational spectrum of the randomly deposited adsorbate molecule is then monitored by Fourier transform reflection absorption infrared spectroscopy in real time. The CO stretching frequency for molecules adsorbed on the (111) terrace of platinum is 2088 cm^{-1}.[26] When the molecule is adsorbed at the step edge, the stretching frequency is 2059 cm^{-1}.[26] The relative integrated intensities of these two bands as a function of time can then be used to extract information about the diffusion of molecular CO across the Pt(111) terraces to their energetically favored adsorption site at the step.

A typical set of vibrational spectra obtained in this manner is illustrated in Figure 5.6 for an initial coverage of $\theta = 0.009$ ML on a Pt[12(111) $\times$ (1$\bar{1}$0)] stepped surface at 117 K.[27] Note that the 2059 cm^{-1} band grows in intensity with time, corresponding to the migration to and adsorption of the CO at step sites. This spectroscopic data can then be converted to CO population at step and terrace sites as a function of time, as shown in Figure 5.7.

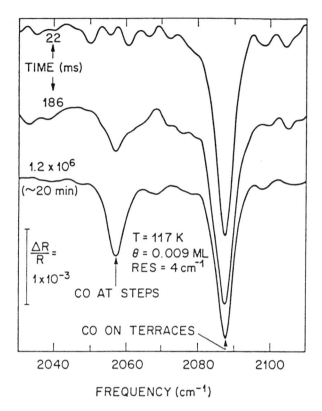

Figure 5.6. Time resolved infrared reflection absorption spectra for CO on a stepped Pt surface, at a coverage of 0.009 ML. After ref. 27.

These measurements have been carried out for two stepped platinum surfaces; one with 12 atom wide terraces, the other with terraces 28 atoms wide. Coverages ranging from 0.004 to 0.013 ML have been examined, and temperature was varied from 100 to 200 K over this range of conditions. For both stepped surfaces, a kinetic model assuming random, uncorrelated site-to-site hopping of the CO molecules across the terrace, was used to fit the diffusional data. Activation energy and preexponential parameters derived from this fitting are $E_{diff} = 4.0 \pm 0.7$ kcal/mol and $A_{diff} = 1.6 \times 10^9$ sec^{-1}. These parameters are somewhat less than previously determined for this system by field emission ($E_{diff} = 14.5$ kcal/mol)[28] or by thermal He scattering ($E_{diff} = 7$ kcal/mol based on an assumed preexponential of 10^{11} sec^{-1}).[29] Comparison of the measured diffusion barrier height with a barrier estimated from the 48 cm^{-1} frequency of the frustrated translation of CO on the Pt(111) surface (5.2 kcal/mol) gives excellent agreement.[30].

This pulsed molecular beam approach, while applied so far to only a single diffusion system, offers great promise for the measurement of adsorbate diffu-

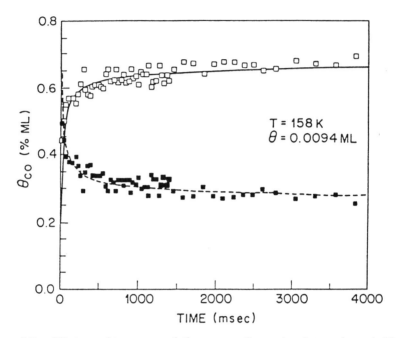

Figure 5.7.   CO step and terrace populations versus time, using data as shown in Figure 5.6. After ref. 27.

sion generally. It probes a length scale intermediate between the long distance laser desorption methods described above, and the very short range techniques such as field emission or scanning tunneling microscopy. It also measures diffusion across defect free areas of the surface avoiding the averaging of terrace, step, and kink sites inherent in the laser desorption method. It can also be used at very low coverages, and can also be applied to modified surfaces or to studies of diffusion in the presence of coadsorbates. It is clearly a complementary method for the determination of surface diffusion dynamics, and is likely to receive much further use in the study of molecular adsorption systems.

5.3 Comments

There are a number of unresolved questions and areas for further research in the investigation of the dynamics of surface diffusion. As we have seen from the examples presented in this chapter, there is a need for a further generalization of experimental methods for probing surface diffusion. In addition to the methods discussed here, the use of scanning probe microscopy to directly monitor surface diffusion seems to be a promising avenue for further work. Scanning tunneling microscopy for metals and semiconductors and scanning force microscopy at atomic resolution for insulators and semiconductors promise a more general approach to the study of molecular diffusion on surfaces. These methods are still very much under development, but they appear to have the generality that is needed to generate a more complete data base for the description of the dynamics of molecular diffusion on surfaces. Because they are localized structural probes, they can be used to monitor the details of diffusion for differing surface structures, near defects, and in the vicinity of steps and edges on the surface. This local nature also means that statistical sampling methods will have to be developed if the diffusion parameters derived from local measurement are to be compared with the results of more macroscopic probes of surface diffusion.

Another underexplored area in the study of surface diffusion is the effect of coadsorbates on molecular surface diffusion. Some work has been carried out exploring the effect of atomic coadsorbates on the diffusion of molecules on surfaces, but very little work has been done to examine the effect of mixed molecular overlayers on the diffusion of a particular molecular species. The diffusion of CO in the presence of hydrogen or hydrocarbon molecules on metal surfaces is a subject that has not been addressed. Further investigation of this sort of model system would help to shed light on the motion of molecular species on surfaces under conditions of catalysis and chemical vapor deposition chemistry, where many species are moving on the surface simultaneously. These studies again will require the development of experimental methods that will be able to monitor the motion of one adsorbate in the presence of another, and that will be generally applicable to the study of molecular diffusion. It is also likely that the dynamic nature of the surface itself will have to be more fully explored in developing a complete understanding of molecular diffusion on surfaces. The surface is not a strictly static template over which the diffusing molecule moves, but rather it is a dynamic ''sea'' supporting the molecule's motion. Cooperative effects, and concerted motions of the surface atoms as the molecule moves across the surface will have to be taken into account, and methods for detecting and understanding this motion will have to be developed.

There is room for work on the theory of surface diffusion as well. Molecular modeling and molecular dynamics simulations offer the prospect of testing specific diffusion mechanisms. Comparison between experimentally derived diffusion parameters and parameters derived from simulations based on model mechanisms will move forward our understanding of the microscopic details of surface diffusion. Computational methods and hardware are becoming available that make this sort of approach feasible, and they will certainly be applied to this aspect of heterogeneous reaction dynamics more frequently in the future.

Future work will also concentrate on the effects of defects on the mechanisms and rates of diffusion on surfaces. As pointed out in the examples above, the presence of steps and defect sites on the surface certainly affect the diffusion of molecules on surfaces. To develop a detailed understanding of the diffusion process, more information will have to be obtained about the structural details of defects and steps. We will have to know what controls step and defect structures and distributions on surfaces. We will have to understand the energetics of adsorption and the dynamics of the interactions of admolecules with these step and defect sites. Scanning probe microscopies and detailed diffraction studies are starting to provide this sort of information, and further work in this area will be welcome.

References

1. G. Ehrlich and K. Stolt, *Annu. Rev. Phys. Chem.* **31**, 603 (1980).

2. R. Gomer, *Rep. Progr. Phys.* **53**, 917 (1990).

3. H. P. Bonzel, U. Breuer, B. Voigtländer, and E. Zeldov, *Surface Sci.* **272**, 10 (1992).

4. H. P. Bonzel, Surface diffusion on metals, Chapter 13 in *Diffusion in Metals and Alloys,* H. Mehrer, ed., Landolt-Bornstein, Berlin, 1990.

5. R. B. Hall and A. M. DeSantolo, *Surface Sci.* **137**, 421 (1984).

6. A. L. Helms, Jr., C.-C. Cho, S. L. Bernasek, and C. W. Draper, The use of LEED for the characterization of surface damage from pulsed laser irradiation, *Materials Res. Soc. Symp. Proc.* **48**, 3 (1985).

7. S. M. George, *Proc. Int. Conf. Lasers* **23**, (1985).

8. J. E. Reutt-Robey, D. J. Doren, Y. J. Chabal, and S. B. Christman, *Phys. Rev. Lett.* **61**, 2778 (1988).

9. C. H. Mak, J. L. Brand, A. A. Deckert, and S. M. George, *J. Chem. Phys.* **85**, 1676 (1986).

10. S. M. George, A. M. DeSantolo, and R. B. Hall, *Surface Sci.* **159**, L425 (1985).

11. R. DiFoggio and R. Gomer, *Phys. Rev.* **B25**, 3490 (1982).

12. H. Conrad, R. Scala, W. Stenzel, and R. Unwin, *J. Chem. Phys.* **81**, 6371 (1984).

13. C. H. Mak and S. M. George, *Chem. Phys. Lett.* **135**, 381 (1987).

14. C. H. Mak, J. L. Brand, B. G. Koehler, and S. M. George, *Surface Sci.* **191**, 108 (1987).

15. C. H. Mak, J. L. Brand, B. G. Koehler, and S. M. George, *Surface Sci.* **188**, 312 (1987).

16. P. Feulner and D. Menzel, *Surface Sci.* **154**, 465 (1985).

17. A. A. Deckert, J. L. Brand, M. V. Arena, and S. M. George, *Surface Sci.* **208**, 441 (1989).

18. H. Pfnür, P. Feulner, and D. Menzel, *J. Chem. Phys.* **79**, 4613 (1983). E. D. Williams, W. H. Weinberg, and A. C. Sobrero, *J. Chem. Phys.* **76**, 1150 (1982).

19. C. H. Mak, B. G. Koehler, J. L. Brand, and S. M. George, *J. Chem. Phys.* **87**, 2340 (1987). J. L. Brand, A. A. Deckert, and S. M. George, *Surface Sci.* **194**, 457 (1988).

20. J. L. Brand, M. V. Arena, A. A. Deckert, and S. M. George, *J. Chem. Phys.* **92**, 5136 (1990).

21. M. V. Arena, A. A. Deckert, J. L. Brand, and S. M. George, *J. Phys. Chem.* **94**, 6792 (1990).

22. R. B. Hall and A. M. DeSantolo, *Surface Sci.* **137**, 421 (1984).

23. E. G. Seebauer, A.C.F. Kong, and L. D. Schmidt, *J. Chem. Phys.* **88**, 6597 (1988).

24. E. G. Seebauer and L. D. Schmidt, *Chem. Phys. Lett.* **123**, 129 (1986).

25. X. D. Zhu, T. Raising, and Y. R. Shen, *Phys. Rev. Lett.* **61**, 2883 (1988).

26. F. M. Leibsle, R. S. Sorbello, and R. G. Greenler, *Surface Sci.* **179**, 101 (1987).

27. J. E. Reutt-Robey, D. J. Doren, Y. J. Chabal and S. B. Christman, *J. Chem. Phys.* **93,** 9113 (1990).

28. R. Lewis and R. Gomer, *Nuevo Cimento* **5,** Suppl. 2, 506 (1967).

29. B. Poelsema, L. K. Verheij, and G. Comsa, *Phys. Rev. Lett.* **49,** 1731 (1982).

30. A. M. Lahee, J. P. Toennies, and C. Woll, *Surface Sci.* **177,** 371 (1986).

CHAPTER

6

Dynamics of Dissociative Adsorption

One of the simplest heterogeneous *reaction* processes that can be considered is the dissociation of a molecular species at a solid surface. This process follows after the energy transfer step considered earlier, and may follow diffusion on the surface. Catalytic dissociative adsorption, where the dissociation products may react further or recombine with one another, but in any case leave the surface eventually, is of primary concern in the examples discussed in this chapter. This is in contrast to oxidative dissociative adsorption, where the adsorption products react with the substrate to make a stable, bound overlayer. The dynamics of catalytic dissociative adsorption have been addressed in several detailed studies, where questions of translational energy and internal energy in the incoming reactant, as well as the structure and composition of the catalytic substrate have been considered. This area has been carefully discussed by Ceyer in a very useful review article.[1] In fact, one of the case histories discussed here is from the Ceyer laboratory, and serves as an excellent example of the degree of detail that can be obtained concerning the dynamics of dissociative adsorption, as well as an example of the clear implications this sort of study has for the understanding of technologically important subjects such as heterogeneous catalysis. The examples considered in this chapter rely heavily on molecular beam approaches for the state preparation and definition of the incident reactant, as well as for product characterization and associated kinetic measurements. The next section provides some background information on these approaches.

6.1 Experimental Methods

Molecular beam methods provide a convenient way to control dynamic variables relevant to the process of dissociative adsorption. In a typical adsorption measurement, a surface would be exposed to an isotropic pressure of a gas phase adsorbate, characterized by its composition and temperature. Molecules from this isotropic gas would be incident on the surface under consideration from all angles, with all orientations, and with a distribution of collision velocities determined by the equilibrium temperature of the gas. Adsorption of the molecules after a finite exposure could then be monitored by spectroscopically examining the surface, or by heating the surface and measuring the quantity of material desorbed into the gas phase.

Molecular beam methods offer the possibility of exploring the detailed dynamics of this process by controlling the angle of incidence, the velocity, the internal energy, and in favorable cases the actual orientation of the incident adsorbate molecule. The technology of nozzle beam sources is quite well developed, and can be used to form a well-collimated, high intensity beam of molecules, with a well-defined translational energy. This translational energy is determined and can be varied by the nozzle stagnation temperature, or by seeding the molecular species in an excess of lighter or heavier noble gas carrier. Translational energies ranging from a few to several tens of kilocalories per mole can be obtained for diatomics and small polyatomics in this way.[2] For a well-collimated beam, the angle of incidence of the molecules with respect to the surface normal, or with respect to particular crystallographic directions in the surface plane, is readily controlled.

For diatomic and polyatomic molecules, the stagnation temperature and expansion conditions also control the internal energy of the molecular adsorbate. This effect can be quite significant depending on the vibrational and rotational energy levels of the molecule. In combination with clever choice of seeding mixtures, the nozzle temperature can be used to more or less independently vary the translational and internal energy of the adsorbate species.[3] The molecular beam approach also allows the possibility of state-selective excitation of the incident adsorbate using various laser or electron excitation techniques. This approach has been used quite successfully to excite significant quantities of NO into the $v = 1$ state for subsequent examination of gas-surface energy transfer.[4] Excitation schemes for a number of other molecular adsorbate species are being developed and applied to studies of all aspects of heterogeneous reaction dynamics.

One of the most exciting developments in this arena is the use of various molecular beam/laser methods to prepare oriented or aligned beams of molecular adsorbate species.[5] These specially prepared molecules can then be used to probe in detail important questions of energy transfer and heterogeneous reaction. For example, the effect of the plane of rotation of an incident molecule (parallel or perpendicular to the surface plane) on its dissociative adsorption probability could be examined. When coupled with a careful characterization and control of the surface structure and properties, a very detailed look at steric effects and the potential energy hypersurface in general becomes feasible. These sorts of experiments are just beginning to be carried out.[6]

At any level of incident beam characterization and control, it is essential that the results of the surface collision process be detectable. For studies of dissociative adsorption, this generally involves examining the substrate surface for evidence of dissociative adsorption products or monitoring the production of subsequent products desorbed into the gas phase. Electron spectroscopy, specifically AES or HREELS methods, can be used to detect and identify the surface bound products of dissociative adsorption. Probes such as TEAS can also be used to monitor dissociative adsorption product coverage on the surface. Mass spectrometry is typically used to monitor and quantify subsequently desorbed product species, either by simple thermal desorption measurements, or using more sophisticated kinetic measurements such as modulated molecular beam reactive scattering (MBRS) approaches. Figure 6.1a illustrates an experimental apparatus using electron spectroscopy and TDS as the primary probes of the dissociative adsorption process. Figure 6.1b shows an example of an instrument that can use TDS, TEAS, or MBRS methods to monitor the dissociative adsorption.

The MBRS method has been described in a number of review papers and research articles.[7] It has been applied to the kinetic study of a wide range of surface reaction processes, including dissociative adsorption,[8] catalytic oxidation reactions,[9] and surface etching reactions.[10] In this method, a beam of molecules (whose incident angle, energy, and orientation can be controlled as described above) is periodically modulated with a known modulation function. The beam is incident on the active surface, and reaction (dissociative adsorption in this case) is monitored by detection of the modulated waveform of the scattered or desorbed product. For simple dissociative adsorption, the detected product might be the recombined dissociated adsorbate, perhaps isotopically labeled to increase signal to background. From the measured product waveform and the known incident waveform, a surface reaction operator is proposed based on a kinetic model for the process. The ability of the proposed kinetic scheme to predict the measured product waveform for various system conditions (incident beam conditions, surface properties, and temperature) provides a measure of the validity of the proposed kinetic model. The measured product intensity also indicates the degree of dissociative adsorption, and when coupled with control of the incident beam parameters gives detailed dynamic information about the process.

An example of the application of MBRS to the study of a simple dissociative adsorption process is illustrated in what follows.[11] Consider a beam of diatomic molecules A_2 of intensity I_0 chopped by a gating function $g(t)$ that is periodic with angular frequency ω. These molecules interact with the surface with sticking probability η and desorb with rate constant k_d.

$$A_{2(g)} + 2S \xrightarrow{\eta I_0} 2[SA_{ads}] \xrightarrow{k_d} 2S + 2A_{(g)} \qquad (6.1)$$

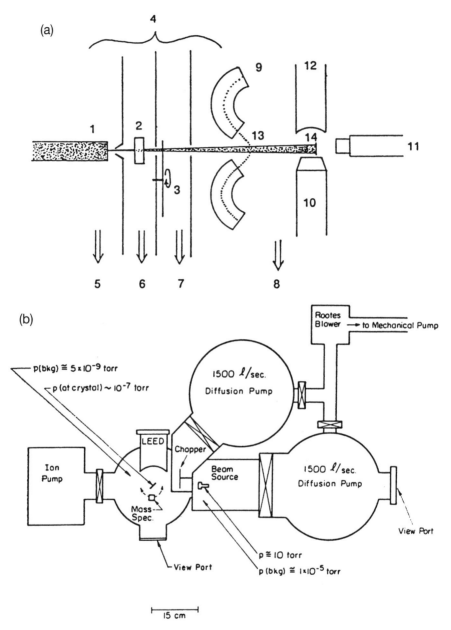

Figure 6.1. (a) Schematic drawing of a molecular beam-UHV apparatus using electron spectroscopy to detect reaction. (1) Nozzle; (2) electronic shutter; (3) chopper; (4) differential pumping stages; (5) to 10-in. diffusion pump; (6) to 6-in. diffusion pump; (7) to 4-in. diffusion pump; (8) to liquid nitrogen trapped 10-in. diffusion pump; (9) HREEL spectrometer; (10) single pass CMA Auger spectrometer; (11) quadrupole mass spectrometer; (12) LEED apparatus; (13, 14) possible sample positions. After ref. 1. (b) Schematic diagram of molecular beam apparatus using mass spectrometer to detect reaction. After ref. 30.

A surface mass balance on adsorbed A atoms gives

$$\frac{d[SA_{ads}]}{dt} = 2\eta I_0 g(t) - k_d [SA_{ads}] \tag{6.2}$$

Substituting a sinusoidal function for $g(t)$ and a trial solution for $[SA_{ads}]$ gives

$$i\omega[SA_{ads}]^* \, e^{i\omega t} = 2\eta I_0 g_1 e^{i\omega t} = k_d[SA_{ads}]^* \, e^{i\omega t} \tag{6.3}$$

Equation (6.3) is solved for $[SA_{ads}]^*$ to give

$$[SA_{ads}]^* = \frac{2\eta I_0 g_1}{k_d + i\omega} \tag{6.4}$$

Writing the complex number in polar form and solving for the rate of desorption $(k_d[SA_{ads}]^*)$ gives

$$k_d[SA_{ads}]^* = \frac{2\eta I_0 g_1 e^{-i\tan^{-1}(\omega/k_d)}}{\sqrt{1 + (\omega/k_d)^2}} \tag{6.5}$$

A reaction product vector can be defined that is the ratio of scattered product signal to incident reactant flux. In the limit of low reaction probabilities this vector is just the ratio of $k_d[SA_{ads}]^*$ to $I_0 g_1$ modified by a phase factor related to the surface residence time of the products.

$$\epsilon = \frac{k_d[SA_{ads}]^*}{I_0 g_1} e^{-i\phi} = \epsilon e^{-i\phi} \tag{6.6}$$

Equation (6.5) can then be written

$$\epsilon e^{-i\phi} = \frac{2\eta e^{-i\tan^{-1}(\omega/k_d)}}{\sqrt{1 + (\omega/k_d)^2}} \tag{6.7}$$

ϵ, the ratio of product to reactant signal, is given by

$$\epsilon = \frac{2\eta}{\sqrt{1 + (\omega/k_d)^2}} \tag{6.8}$$

and ϕ, the phase difference between product and reactant signals, is given by

$$\phi = \tan^{-1}(\omega/k_d) \tag{6.9}$$

By observing the amplitude and phase of the product and reactant signals as a function of chopping frequency ω, k_d can be determined by a plot of $\tan \phi$ vs ω and η can be determined by a plot of $1/\epsilon^2$ vs ω^2. Determinations of k_d at several surface temperatures can give the activation energy and preexponential factor for an Arrhenius-type rate equation. Similar analyses of more complex surface reaction models, including series, parallel, and combination series–parallel models, enable the experimenter to extract kinetic parameters and to choose appropriate surface reaction mechanisms.

In the above analysis, a sinusoidal gating function was employed. For first-order surface processes a sinusoidal gating function is allowed in the analysis regardless of the actual waveform of the incident beam due to the lack of information in higher order Fourier components. For processes other than first order the analysis becomes more complex and must take account of the actual gating function waveform. Several examples of non-first-order processes have been treated by Olander.[12] Work by Foxon et al.,[13] and Sawin and Merrill[14] has shown the advantage of collecting the entire waveform and using fast Fourier transform techniques to extract the kinetic information, even for first-order processes. The advantage to this method is in the increased amount of data available from a single experiment, as the higher harmonics of the signal give information at higher frequencies simultaneously. This method is also directly applicable to non-first-order processes without the more complex analysis necessary in lock in detection methods.

6.2 Illustrative Examples

The examples discussed in this chapter draw on these very powerful methods, as well as others, to provide an increasingly more detailed view of the process of dissociative adsorption. While there are a number of studies in the literature that illustrate the points described here, these three examples are especially useful due to their completeness and rather clear connection to important technological questions. The specific examples to be discussed are the dissociative adsorption of N_2 on Fe(111), the dissociative adsorption of CH_4 on Ni(111), and the dissociative adsorption of H_2 on Pt(111).

6.2.1 Dissociative Adsorption of N_2 on Fe(111)

The dissociative chemisorption of nitrogen on iron has been extensively studied over many years. Several detailed studies have indicated that dissociative adsorption of N_2 on iron is the rate-limiting step for the subsequent production of NH_3 by iron-based catalysts. Careful equilibrium adsorption studies using well-characterized single crystal surfaces have provided a great deal of background information on this adsorption system. Considering the Fe(111) substrate in particular, which has been shown to be the most active for N_2 dissociative adsorption and subsequent ammonia formation, a number of points bearing on the dynamics of dissociative adsorption should be mentioned.[15]

The dissociative adsorption probability for ambient N_2 on Fe(111) is quite low. It has been measured to be 10^{-7}–10^{-6}, significantly less than the dissociative adsorption probability for N_2 on other transition metal surfaces. Even so, the dissociative adsorption probability on Fe(111) is significantly greater than on the Fe(100) or Fe(110) surfaces; Fe(111) is the most active low index face.

Dissociative adsorption is found to occur by way of a molecular precursor state on the Fe(111) surface. Vibrational spectroscopy and photoelectron spectroscopy have identified this precursor as a π-bonded (side-on bonded) molecular species, with a weakened N–N bond. This α adsorption state has a ν(N–N) vibrational frequency of 1490 cm^{-1}, and is the direct precursor to dissociation. An additional molecularly bound adstate has been identified, the γ-N$_2$ state, which appears to be end-on bound, and is a precursor to the α state. The γ-N$_2$ and α-N$_2$ states are separated by an activation barrier of about 10 kcal/mol, as are the α-N$_2$ and the dissociatively adsorbed β state.

Given this background information, molecular beam studies of the dynamics of this dissociative adsorption process can be discussed. The results of such a study are, in fact, quite dramatic. In this example, a beam of N$_2$ molecules of variable translational energy was incident on a clean, well-characterized Fe(111) single crystal substrate in an ultrahigh vacuum apparatus.[16] The translational energy of the incident beam was varied from less than 0.5 eV to greater than 4 eV by varying the nozzle temperature from 300 to 2000 K and by seeding the nitrogen in H$_2$ or He. The dissociative adsorption probability was obtained by measuring the nitrogen coverage on the Fe(111) surface as a function of exposure to the beam using Auger electron spectroscopy. The initial slope of these coverage versus exposure plots provides the relative dissociative adsorption probability at that incident kinetic energy. Absolute dissociative adsorption probability η_0 was determined by the method of King and Wells[17] at high incident

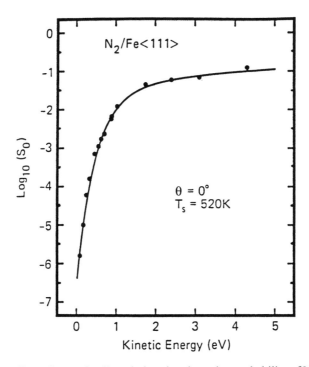

Figure 6.2. Effect of E_i on the dissociative chemisorption probability of N$_2$ on Fe(111) at θ_i-0° and T_s = 525 K. After ref. 15.

energies, and used to calibrate the relative measurements obtained by Auger spectroscopy.

The dramatic results of this measurement are shown in Figure 6.2. The low dissociative adsorption probability characteristic of adsorption at low collision energies (ambient conditions) increases over five orders of magnitude as the incident translational energy is increased. The data shown here correspond to normal incidence measurements, and a surface temperature of 520 K. If direct dissociative adsorption is occurring, this observation is perhaps not so surprising. Direct dissociative adsorption is often activated, that is, an energy barrier must be surmounted along the dissociation coordinate. Previous studies have indicated that the dissociative adsorption process is not activated in this system, and there is clear evidence that it proceeds via the π-bonded molecular precursor discussed earlier.

Previous work has also shown a strong surface temperature dependence for the dissociative adsorption process. The dissociative adsorption probability increases with decreasing temperature, as would be expected for competing rates of dissociation and desorption on a molecular precursor state. When the surface temperature dependence for the dissociative adsorption probability is investigated for various incident translational energies, similar behavior is observed. Figure 6.3 shows the observed increase in η_0 with decreasing surface temperature, for an incident translational energy of 1.05 eV. This behavior is consistent with dissociative adsorption via a precursor, and from the slope of this plot of

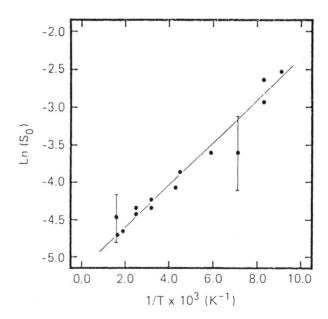

Figure 6.3.   Effect of surface temperature on the dissociative chemisorption probability of N_2 on Fe(111) at normal incidence and at a kinetic energy of 1.05 eV. After ref. 15.

η_0 versus $1/T_s$, provides an activation barrier for dissociation from the precursor state. Similar behavior is seen over the whole range of incident translational energies. These observations provide clear evidence for an activation barrier (actually a distribution of barriers) to access the molecular precursor to dissociative adsorption for N_2 on Fe(111). Translational energy is effective in overcoming this barrier, and increasing the population of the intermediate π-bonded α-N_2 state. Dissociative adsorption from the translationally accessed precursor state is similar in temperature dependence and saturation N-atom coverage to adsorption from ambient gas through this precursor.

Extensions of this work to consider the effect of vibrational energy on η_0 have been made.[18] By using a range of nozzle temperatures and seeding conditions, η_0 was measured for beams with differing amounts of vibrational energy in the incident N_2, but with the same translational energy. It was found that vibrational energy enhanced the dissociative adsorption probability, but not as effectively as an equivalent amount of translational energy. Vibrational energy was estimated to be about 55% as effective as translational energy in accessing the α-N_2 precursor state.

6.2.2 Dissociative Adsorption of CH_4 on Ni(111)

The dissociative adsorption of methane on the nickel surface exhibits an interesting behavior. Estimates of the adsorption probability indicate a very low dissociative adsorption probability, less than 1×10^{-9} for clean, well-characterized single crystal surfaces of nickel exposed to methane at low pressures.[19] Under the conditions of typical UHV adsorption studies, methane is not observed to adsorb dissociatively on the nickel surface. However, commercial catalysts based on nickel are used for the steam reforming of natural gas,[20] whose first reaction step must be the activation of the C–H bond and the dissociative adsorption of methane. Measurable rates of dissociative adsorption have been observed for a Ni(111) crystal exposed to very high pressures of methane,[21] since the higher flux of molecules at higher pressures overcomes the very low dissociative adsorption probability. This observed behavior suggests the presence of an activation barrier along the reaction coordinate for dissociative adsorption of methane on the nickel surface.

In a series of very elegant experiments, this hypothesis has been tested by Ceyer and co-workers.[22] They have combined the molecular beam methods described earlier in this chapter with the use of electron spectroscopic identification of the dissociative adsorption products on a well-characterized nickel single crystal substrate, to probe this activated dissociative adsorption process in detail. The basic approach is to impinge a beam of methane molecules with variable translational and vibrational energy on a clean Ni(111) surface, and to monitor the dissociative adsorption process by monitoring the coverage of carbon on the surface using AES. The carbon coverage, coupled with a knowledge of the flux of methane molecules to the surface at a particular beam energy, provides a measure of the dissociative adsorption probability as a function of the methane translational energy. The results of such a measurement are shown in Figure 6.4. The figure clearly shows a threshold translational energy, below which the dissociation probability is not measurable. Above this energy, the adsorption probability increases exponentially with increasing energy normal to the surface.

The translational energy of the incident beam was varied in this study by varying the nozzle temperature of a 1% methane in helium mixture over the range of 640 to 830 K. The angle of incidence was also varied from normal to the surface to around 40° away from the surface normal. The dissociation probability for this range of conditions was found to vary with the normal component of the methane incident kinetic energy and not with the total translational energy. The dissociation probability for deuterated methane was also determined in this fashion and is shown in Figure 6.4 as well. Similar behavior is observed, with a threshold evident for the onset of dissociation, and an exponential increase in dissociation probability with the normal component of the energy of the incident molecule. The absolute value of the dissociation probability, however, is about an order of magnitude smaller in the case of the deuterated molecule.

The effect of vibrational energy on the dissociation process was also examined.[23] By varying the nozzle temperature and the amount of methane in the helium carrier gas, the vibrational energy of the methane molecule could be

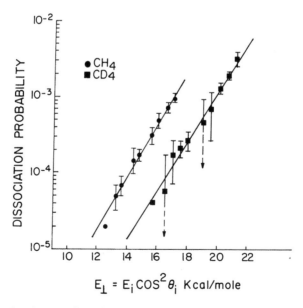

$$E_{\perp} = E_i \cos^2 \theta_i \quad \text{Kcal/mole}$$

Figure 6.4. The absolute dissociation probability of CH_4 and CD_4 as a function of the normal component of translational energy. The lines are linear least squares fit to the data and the error bars are 95% confidence limits of a series of 6–8 measurements for CH_4 and of a series of three measurements for CD_4. After ref. 23.

varied for a fixed incident translational energy. Using this approach, the vibrational energy was varied from 0.96 to 1.45 kcal/mol for translational energies in the range of 14 to 17 kcal/mol. These measurements indicated that vibrational energy in this range was approximately equally effective in promoting dissociative adsorption of methane on the nickel surface as was translational energy. That is, a given change in vibrational energy produced the same variation in dissociation probability as would result from that same change in translational energy with fixed vibrational energy. Vibrational energy by itself is not effective in promoting dissociation, as the accessible vibrational energy in these experiments is not sufficient to surpass the threshold indicated in the data of Figure 6.4.

These experiments also provided direct spectroscopic identification of the dissociative adsorption products on the nickel surface.[24] High resolution electron energy loss spectroscopy was used to examine the surface following exposure to the translationally excited methane beam. The observed spectrum was consistently assigned as being due to adsorbed methyl radicals on the surface. These methyl radicals are stable at surface temperatures up to 150 K. Above this temperature, reaction occurs to form C_2H_2, which dissociates to give adsorbed CH, H, and eventually desorbed hydrogen and adsorbed carbon.

It is clear from the observations in this example that there is a barrier to dissociative adsorption of methane on the Ni(111) surface. Translational energy normal to the surface is effective in overcoming this barrier. Vibrational energy in the methane molecule is also seen to be effective in promoting dissociative adsorption, and appears to be at least as effective as translational energy in surmounting this barrier. There is a threshold of about 12 kcal/mol for dissociative adsorption, and the observed dissociation probability for deuterated methane is an order of magnitude lower than the normal methane molecule. A physically attractive model has been proposed by Ceyer and co-workers to account for these specific dynamic observations.[25]

This model is essentially a molecular deformation model, which assumes a transition state for the dissociative adsorption process that distorts the methane molecule from its normal tetrahedral configuration on collision with the surface. Since both translational energy and vibrational energy appear to be equally effective in promoting dissociative adsorption, translational and vibrational excitation must result in the same motion of the relevant nuclei along the reaction coordinate. This suggests that the role of the collision of the methane molecule with the repulsive wall of the surface interaction potential is to convert the translational energy of the molecule into vibrational motion along the reaction coordinate. This motion is associated with the ν_2 and ν_4 vibrational modes in the methane molecule, and these modes are also the modes excited by heating the nozzle source in the vibrational energy dependence measurements. This conversion of translational energy to vibrational energy in the methane molecule results in a distortion of the molecule from its tetrahedral configuration, allowing the hydrogen atoms to move out from between the surface and the incoming carbon atom of the methane molecule. The more exposed carbon atom can then interact with the nickel surface, forming a Ni–C and Ni–H bond and breaking a C–H bond. In this sense, the barrier to dissociative adsorption is the energy needed to distort the molecule into the configuration that can interact with the metal surface. This distortion energy has been calculated by Ceyer and co-workers, using simple force field calculations, and assuming a pyramidal configuration for the transition state, with the carbon and three hydrogens all in the same plane. Using a harmonic force field this energy barrier is calculated to be about 16 kcal/mol, similar to the threshold energy for the dissociative adsorption determined from the molecular beam experiments.

This physical model can also explain the difference in dissociative adsorption probability between methane and deuterated methane. Experiments indicate that the dissociative adsorption probability of methane is an order of magnitude higher than the dissociative adsorption probability for deuterated methane with the same kinetic energy normal to the surface. This difference can not be explained by a kinetic isotope effect due to the difference in the zero point energies of the two molecules. Rather, the model set forth by Ceyer suggests a tunneling step along the reaction coordinate for C–H bond cleavage. Tunneling rates could easily be an order of magnitude higher for light hydrogen atom motion along this coordinate compared to deuterium atom motion along the same coordinate. This combination of molecular deformation to bring the carbon atom near the metal surface with tunneling along the C–H bond cleavage reaction coordinate accounts for the major features of the observations in this study of the dissociative adsorption of methane on the Ni(111) surface. These major features are the normal kinetic energy scaling of the dissociative adsorption probability, the 16 kcal/mol threshold for the onset of dissociation, the observed equivalence of translational and vibrational energy in the dissociation process, and the large isotope effect in comparison of methane and deuterated methane dissociative adsorption probability.

6.2.3 Dissociative Adsorption of H_2 on Pt(111) and Stepped Pt Surfaces

The dissociative adsorption of hydrogen on platinum is technologically a very important surface process. Hydrogenation catalysis, water electrolysis, platinum-catalyzed reforming reactions, and hydrogenolysis reactions all involve the dissociation, diffusion, and recombination of hydrogen on the platinum surface. The dissociation reaction on well-characterized platinum surfaces has been widely studied, and has served as a model diatomic dissociation system. Early thermal desorption studies and isotope exchange measurements suggested that hydrogen readily dissociates on platinum, even at low temperatures.[26] Equilibrium (thermal desorption) experiments carried out on low index surfaces such as the Pt(111) surface supported this conclusion as well;[27] the platinum surface readily dissociates molecular hydrogen.

Molecular beam scattering experiments carried out in the early 1970's began to call this general conclusion into question. Direct measurement of the hydrogen–deuterium exchange reaction rate on the Pt(111) surface was compared with the exchange rate on a stepped platinum [Pt(997)] surface.[28] It was found that the stepped surface was much more active for the exchange reaction than was the low index (111) surface. It was concluded that the step sites on the (997) surface were the active sites for the dissociation reaction, and that in the limit of no step sites on the platinum surface there would be very little hydrogen dissociation. This conclusion was further supported by more extensive modulated molecular beam scattering studies of the mechanism of the exchange reaction.[29] These studies indicated that the dissociation occurred by way of a branched mechanism. One branch was identified with "direct" dissociation of the incident hydrogen molecule in the vicinity of the step edge, and the other branch was identified with adsorption of the hydrogen molecule on the (111) terrace regions of the surface followed by diffusion to the step site where the dissociation took place.

Further molecular beam scattering studies by Somorjai and co-workers[30] examined the effect of these step sites on the dissociative adsorption probability in more detail. In these investigations, the distinct advantage of the molecular beam scattering approach was put to use. The beam of reactant deuterium was brought incident to the stepped surface either parallel or perpendicular to the step direction, and the rate of exchange to produce HD monitored as this incident orientation was changed. When the beam was incident perpendicular to the step edges, it could also be brought in from the "downstairs" or "upstairs" direction with respect to the steps. Differences in HD product formation rate were observed as illustrated in Figure 6.5.

As can be seen, the HD production on the Pt(332) stepped surface is independent of the in-plane angle of incidence when the reactant beam is parallel to the step edges. When the surface is rotated 90°, and the reactant beam is brought in perpendicular to the step edges, there is a strong dependence on the in-plane angle of incidence. The production rate varies from a value below that of the parallel orientation rate for molecules incident from the top side of the step, to a value that is well above the parallel orientation for the reactant beam incident from the "downstairs," or open side of the step. The same investigators examined the reaction on the Pt(111) surface and found a much lower reaction rate, consistent with previous measurements. As seen in Figure 6.5, the angular dependence of the reaction rate on the Pt(111) surface showed a slight peaking at the normal to the surface, suggesting that there is a small activation barrier to dissociative adsorption of H_2 on this surface. No incident azimuthal angular dependence was observed for the (111). Very similar behavior was observed for the other stepped surface investigated in this study, the Pt(553) surface, i.e., a strong azimuthal dependence for the beam incident parallel or perpendicular to the step edges, and a higher reaction probability when the reactant beam is incident on the open side of the step edge. These results suggest that dissociation of the H_2 molecule occurs at the step edge directly on impact.

These workers further examined the exchange reaction using the modulated

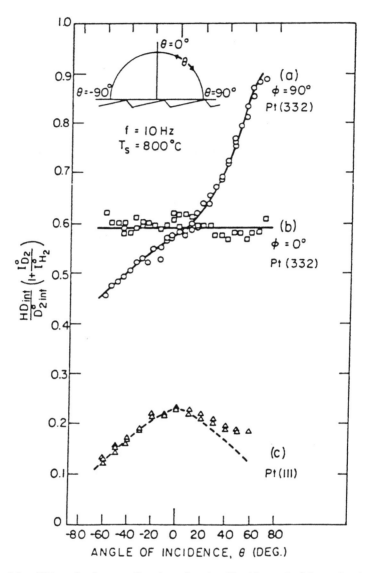

Figure 6.5. HD production as a function of angle of incidence θ of the molecular beam normalized to the incident D_2 intensity. (a) Pt(332) {or Pt(S)—[6(111) × (111)]} surface, with the step edges perpendicular to the incident beam ($\phi = \pm 90°$); (b) Pt(332), where the projection of the beam on the surface is parallel to the step edges ($\phi = 0°$). (c) Pt(111) surface.

beam approach described earlier in this chapter.[31] A mixed H_2–D_2 beam was modulated over the frequency range from 2 to 200 Hz, and the exchange reaction was monitored by detecting HD for various final scattering angles over the surface temperature range from 25 to 800°C. Both the Pt(111) and the stepped Pt(332) surface were studied. Product HD amplitude and product phase lag were monitored over this range of reaction conditions, and the data were best described by a branched reaction mechanism, in agreement with earlier modulated molecular beam scattering work for this reaction on the Pt(997) surface. Detailed analysis of the reaction mechanism and fitting to the experimental data indicate that the adsorption and dissociation of H_2 on the Pt(111) surface are activated processes, while on the stepped surface there is no activation barrier. The angular anisotropy observed in the earlier work is maintained in these more detailed studies, down to surface temperatures as low as 200°C. After dissociation, the recombination and desorption of the hydrogenic molecules appear to follow the same mechanism for the flat and the stepped surfaces. One of the parallel branches is operative over the entire temperature range studied, and has an activation energy of 13.0 $\pm$ 0.4 kcal/mol, and psuedo-first-order preexponential factor of 8×10^4 sec^{-1} on the stepped surface and a slightly higher activation energy on the (111) surface. The other branch appears to operate above 300°C, but the activation energy and preexponential factor of this branch could not be uniquely determined.

The dissociative adsorption of H_2 on the Pt(111) surface has been revisited more recently by Poelsema and co-workers.[32,33] Their studies have used thermal energy atom scattering (TEAS) and molecular beam scattering methods to explore the interaction of hydrogen with the Pt(111) surface in great detail, addressing the controversy generated by earlier work mentioned above, which suggested the importance of step defect sites on the dissociative adsorption process. Using a thermal energy helium beam to probe the coverage of atomic hydrogen on the platinum surface as a function of hydrogen exposure, Poelsema and co-workers[32] investigated the adsorption over a wide temperature range for a series of Pt(111) surfaces with increasing step densities. Step defects were introduced into a nearly perfect Pt(111) surface by in situ high temperature ion bombardment. Helium scattering was used to quantify the defect density prior to hydrogen exposure. This method for preparing characterizable defect densities had been developed in the Jülich laboratories in previous studies of the mechanism of ion bombardment etching of the Pt(111) surface,[34] and takes advantage of the carefully prepared nearly defect free Pt(111) surfaces prepared at this institute.[35] The results of the TEAS studies of this adsorption system are summarized in Figure 6.6. The dissociative sticking probability for H_2 on platinum surfaces with varying defect densities is plotted versus hydrogen coverage at a surface temperature of 155 K in this figure. There are clearly two separate regions in the plot, one that is very defect density dependent and independent of hydrogen coverage, and another that is less dependent on defect density, but exponentially dependent on hydrogen coverage. Based on these data, Poelsema and Comsa proposed that dissociation occurs on the platinum surface by way of an

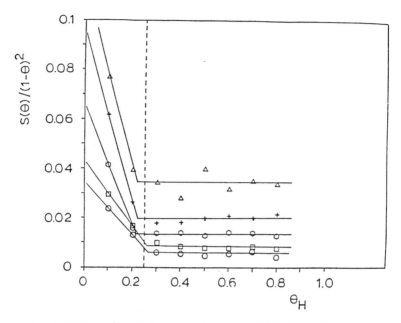

Figure 6.6. $S(\theta)/(1 - \theta)^2$ vs θ_H for adsorption at $T_s = 160$ K on surfaces with a variety of defect densities. Two adsorption channels are visible, and the sticking probability in both channels increases in the presence of defects. The curves shown are for surfaces with approximate defect densities of 0.1% ($\circ$), 1% ($\square$), 2% ($\circ$), 3% ($+$), and 4% ($\triangle$).

adsorbed precursor, and exclusively at step sites on the surface. These step sites can be reached either by diffusion of the precursor molecule from its initially adsorbed site on the Pt(111) terrace, or by direct adsorption at or near the step site. This model is in essential agreement with that proposed by Bernasek et al.[28] based on the molecular beam H_2–D_2 exchange studies described earlier in this chapter.

Again using the H_2–D_2 exchange reaction to monitor the dissociative adsorption of hydrogen on platinum, Verheij et al.[33] carefully investigated this adsorption and desorption process. A molecular beam scattering apparatus with a rotatable, modulated source, a rotatable sample in an ultrahigh vacuum scattering chamber, and a fixed, differentially pumped detector chamber was used to carry out this study. A psuedo-random chopper in the detector chamber was used to obtain TOF information about the scattered reaction products. Angle-dependent sticking probabilities, HD product angular distributions, and velocity distributions of the product molecules were all measured over the surface temperature range from 570 to 1200 K. A H_2 nozzle beam was directed onto the platinum surface, which was exposed to a background pressure of D_2. Product HD is monitored with the detector, as is the scattered unreacted H_2.

The angular distribution of the desorbed product HD is reasonably well fit by a $\cos^5 \theta_f$ distribution. Deviations from this $\cos^5$ behavior for large scattering angles suggested an additional channel for the desorption process, with an overall angular dependence of the HD desorption flux well described by $I_{HD} = A \cdot \cos^5 (\theta_f) + (1 - A)\cos(\theta_f)$, with $A = 0.85$. Velocity distributions of the product HD molecules measured using the psuedo-random TOF chopper indicate that the mean desorption energy of the HD product varies with surface temperature as $E_d = 4kT_s$, and stays relatively constant with scattering angle from the surface normal out to about 45°. For scattering angles larger than 45°, the mean desorption energy drops off quickly. Careful measurements of the dissociative adsorption probability as a function of incident angle and surface temperature and application of the principle of detailed balance suggest a two-channel adsorption mechanism described by $S_a = S_1 E_\perp^2 + S_2$. One channel is strongly (quadratically) dependent on the perpendicular energy of the incident hydrogen molecule. At low surface temperatures, or for high incident energies this channel dominates the adsorption. At higher surface temperatures, the second energy-independent term becomes significant. This second channel is identified with a molecular precursor state; this is clearly consistent with the increase in S_2 with surface temperature, and its decrease with increasing incident energy. The energy-dependent channel exhibits a decrease in magnitude with increasing surface temperature. This observation is inconsistent with an activated adsorption process, as suggested by Salmeron and others,[30,31] but it can be explained in the context of the precursor model suggested in the TEAS studies described above. The dissociation of the hydrogen molecule then takes place at step sites as suggested by Poelsema and Comsa,[32] and earlier by Bernasek and Somorjai.[28,29]

Our discussion of the final example of this chapter will close with a brief mention of recent computer simulation studies of this interesting adsorption process. Classical Monte Carlo simulations of the hydrogen dissociation process based on the results described above have shown that the behavior of the system can be classified into three surface temperature regions.[36] The conclusions of the studies described above, that dissociation occurs only at step sites, and that the observed dissociative sticking coefficient results from two competing channels (precursor adsorption on the terrace followed by diffusion to the step site, and adsorption at or near the step site directly), are incorporated into this simulation. The results of the simulation are quite consistent with this picture of the dissociation, and further suggest that there are three separate temperature regimes in this dissociation process. In the lowest temperature regime, the dissociation process is dominated by the presence of the cluster size distribution of vacant adsorption sites. As temperature increases, these clusters interact with nearest-neighbor adsorbed H atoms, and at the highest temperatures, direct interaction of the incident molecule with a step site dominates. Rather quantitative agreement between the simulation and the sticking coefficient data of Poelsema and Comsa[32] is obtained, even though this simulation ignores possible quantum mechanical effects or the possibility of cooperativity in diffusion in the precursor state.

6.3 Comments

Two general areas for further work are suggested by the illustrative examples of this chapter. The first general area has been clearly hinted at in the work already described. In each of the dissociative adsorption studies discussed here, attempts were made to control the initial state of the dissociating molecule. In each case incident translational energy and incident angle were carefully controlled, and the effects of these experimental parameters on the mechanism and dynamics of the dissociative adsorption process were monitored. In each case the effect of internal energy on the dissociative adsorption also appeared. However, in the case of internal energy, the degree of control has not been as complete as is the case for translational energy. As was seen for both the N_2 and CH_4 studies, some control of vibrational energy is available by a combination of nozzle heating and beam seeding. This gives an average internal energy, however, and does not address the question of state-specific internal energy effects. Clearly an area for future work would be the application of one of the growing number of laser methods, or the development of a new method for the efficient state selective excitation of incident reactants in the cases discussed here. State specific vibrational excitation of N_2 and H_2 have been accomplished,[37] and application of these methods to the study of the dissociative adsorption of N_2 or H_2 would be a welcome dynamic study. Similarly, there have recently been developed efficient means for exciting C–H stretching motions in hydrocarbons,[37] and the direct effect of C–H excitation on the dissociative adsorption of methane would be interesting to know. In all of these cases, the problem of course is to obtain enough of the incident molecules in the specific internally excited state to allow an unambiguous determination of the effect of vibrational or rotational excitation on the dissociation to be made.

The second clear area demanding further attention is in the theoretical and computational interpretation of the results of the sorts of measurements described in this chapter. This has been hinted at in the discussion of the molecular deformation model developed by Ceyer[25] to explain the methane dissociation results described here. As the quality of heterogeneous reaction dynamics data improves, and the number of interesting systems increases, the need for more detailed modeling studies will also increase. The use of classical trajectory calculations to model more and more complex systems, the development of more realistic semiempirical potential energy surfaces, and the eventual development of first principles interaction surfaces for heterogeneous reaction processes will become more widespread. As calculation hardware develops, the complex multidimensional, many body problem of surface reaction dynamics will begin to become tractable. As the quality of the experimental data continues to improve, the theoretical tools will become available to handle these problems. This is clearly an area where experimental and calculational developments will have to take place in concert.

References

1. S. T. Ceyer, *Annu. Rev. Phys. Chem.* **39**, 479 (1988).

2. G. Scoles, ed., *Atomic and Molecular Beam Methods,* Vol 1, Oxford University Press, Oxford, 1988.

3. N. Abauf, J. Anderson, R. Andres, J. Fenn, and D. Marsden, *Science* **155**, 997 (1967).

4. J. Misewich, H. Zacharias, and M.M.T. Loy, *Phys. Rev. Lett.* **55**, 1919 (1985).

5. S. R. Gandhi and R. B. Bernstein, *J. Chem. Phys.* **93**, 4024 (1990).

6. S. I. Ionov, M. E. Lavilla, and R. B. Bernstein, *J. Chem. Phys.* **93**, 7416 (1990).

7. M. P. D'Evelyn and R. J. Madix, *Surface Sci. Rep.* **3**, 413 (1984). J. A. Schwarz and R. J. Madix, *Surface Sci.* **46**, 317 (1974).

8. G. E. Gdowski, J. A. Fair, and R. J. Madix, *Surface Sci.* **127**, 541 (1983).

9. L. S. Brown and S. J. Sibener, *J. Chem. Phys.* **89**, 1163 (1988).

10. R. J. Madix, R. Parks, A. A. Susu, and J. A. Schwarz, *Surface Sci.* **24**, 288 (1971).

11. S. L. Bernasek, Ph.D. dissertation, University of California, Berkeley (1975). R. H. Jones, Ph.D. dissertation, University of California, Berkeley, (1971).

12. D. R. Olander, in *The Structure and Chemistry of Solid Surfaces,* G. A. Somorjai, ed., John Wiley, New York, 1968, p. 45-1.

13. C. T. Foxon, M. R. Boudry, and B. A. Joyce, *Surface Sci.* **44**, 69 (1994).

14. H. H. Sawin and R. P. Merrill, *J. Vac. Sci. Technol.* **19**, 40 (1981).

15. (a) G. Ertl, *Catal. Rev. Sci. Eng.* **21**, 2101 (1984); (b) G. Ertl, *J. Vac. Sci. Technol.* **11**, 1247 (1984).

16. C. T. Rettner and H. Stein, *Phys. Rev. Lett.* **59**, 2768 (1987).

17. D. A. King and M. G. Wells, *Surface Sci.* **29**, 454 (1972).

18. H. Stein and C. T. Rettner, *J. Chem. Phys.* **87**, 770 (1987).

19. F. C. Schouten, O.L.J. Gijzeman, and G. A. Bootsma, *Bull. Soc. Chim. Belg.* **88**, 541 (1979).

20. J. P. Van Hook, *Catal. Rev. Sci. Eng.* **21**, 1 (1980).

21. T. P. Beebe J., D. W. Goodman, B. D. Kay, and J. T. Yates, Jr., *J. Chem. Phys.* **87**, 2305 (1987).

22. S. T. Ceyer, J. D. Beckerle, M. B. Lee, S. L. Tang, Q. Y. Yang, and M. A. Hines, *J. Vac. Sci. Technol. A* **5**, 501 (1987).

23. M. B. Lee, Q. Y. Yang, and S. T. Ceyer, *J. Chem. Phys.* **87**, 2724 (1987).

24. M. B. Lee, Q. Y. Yang, and S. T. Ceyer, *J. Chem. Phys.* **85**, 1693 (1986).

25. J. D. Beckerle, Q. Y. Yang, A. D. Johnson, and S. T. Ceyer, *J. Chem. Phys.* **86**, 7236 (1987).

26. K. Christmann and G. Ertl, *Surface Sci.* **60**, 365 (1976).

27. K. E. Lu and R. R. Rye, *Surface Sci.* **45**, 67 (1974).

28. S. L. Bernasek, W. J. Siekhaus, and G. A. Somorjai, *Phys. Rev. Lett.* **30**, 1202 (1973).

29. S. L. Bernasek and G. A. Somorjai, *J. Chem. Phys.* **62**, 3149 (1975).

30. M. Salmerón, R. J. Gale, and G. A. Somorjai, *J. Chem. Phys.* **70**, 2807 (1979).

31. M. Salmerón, R. J. Gale, and G. A. Somorjai, *J. Chem. Phys.* **67**, 5324 (1977).

32. B. Poelsema and G. Comsa, *Scattering of Thermal Energy Atoms from Disordered Surfaces,* Vol. 15 of Springer Tracts in Modern Physics, Springer, Berlin, 1989.

33. L. K. Verheij, M. B. Hugenschmidt, A. B. Anton, B. Poelsema, and G. Comsa, *Surface Sci.* **210**, 1 (1989).

34. B. Poelsema, L. K. Verheij, and G. Comsa, *Phys. Rev. Lett.* **53**, 2500 (1984).

35. U. Linke and B. Poelsema, *J. Phys. E.* **18**, 26. (1985).

36. H.-P. Kaukonen and R. M. Nieminen, *Surface Sci.* **247**, 43 (1991).

37. H. Zacharias, *Int. J. Mod. Phys. B* **4**, 45 (1990).

CHAPTER

7

Atom Recombination Dynamics

7.1 Introduction

Atom recombination on surfaces is another elementary surface reaction process that has been relatively well studied. The catalyzed recombination of adsorbed atomic species is the reverse of the process discussed in detail in Chapter 6, and has provided some of the most extensive and interesting experimental information about heterogeneous reaction dynamics. In this chapter, examples of the catalytic recombination reaction will be discussed, with emphasis on experimental studies of the dynamics of the recombination reaction as reflected in the angular, velocity, and internal state distribution of the desorbing product molecule. This has been a rather active area of research over the past several years, and there are a number of review papers in the literature that discuss various aspects of this process. Two particularly useful reviews are the one by Comsa and David,[1] entitled "Dynamical parameters of desorbing molecules," and the one by Zacharias,[2] entitled "Laser spectroscopy of dynamical surface processes." In addition to these very clear reviews, which concentrate on the basic description of the desorption process, and the use of laser methods to probe desorption dynamics, there are a number of other review papers that are listed in the references at the end of this chapter.[3]

In this chapter, we will first discuss two experimental techniques that have proven to be crucial to the study of the dynamics of recombinative desorption. We will then describe three example studies that address the questions of the dynamics of recombinative desorption. The first example, the recombination of N atoms on the iron surface, was the first study to identify internal state excitation

in the products of a surface reaction. The second and third examples both deal
with the recombination of H atoms on well-characterized surfaces. In the first
instance the recombination reaction takes place on copper, a surface that exhibits
an activation barrier for the dissociative adsorption of molecular hydrogen. This
recombinative desorption has been well studied, with experimental information
ranging from angular distributions, to measures of the velocity of the desorbing
hydrogen, to elegant studies of the internal energy of the desorbing species. The
last example, H atom recombination on palladium, has also been probed in detail.
This system is particularly interesting due to the high solubility of atomic hydrogen in the bulk of palladium, and the effect that this solubility has on the recombination dynamics.

7.2 Experimental Methods

Two specific experimental methods will be described in this section. The first is
electron beam-induced fluorescence (EBIF),[4] which was in fact the first method
to be used to determine the internal energy of desorbing molecules following
surface-catalyzed recombination. The second, more general approach is termed
resonantly enhanced multiphoton ionization (REMPI or just MPI).[5] Each of these
methods will be discussed briefly, and references provided to more detailed
descriptions in the literature.

Electron beam-induced fluorescence (EBIF) has been applied almost exclusively to the study of nitrogen, although other diatomic and polyatomic molecules can in principle be investigated using this method. In EBIF the molecule
under study, nitrogen in this case, is exposed to a high-voltage, high-current
electron beam. Nitrogen molecules in the electronic ground state are ionized and
electronically excited to the N_2^+ $B^2\Sigma_u^+$ excited state by electron collision. These
electronically excited ions then fluoresce to the N_2^+ $X^2\Sigma_g^+$ state. For nitrogen,
the electron excitation cross section and the fluorescence yield are both quite
high, making this method a sensitive probe of the density of N_2 molecules in
the electron beam if the total fluorescence signal is collected.

If the fluorescence signal is focused on a monochromator and dispersed, the
intensities of emission from individual vibrational bands, and from the rotationally resolved spectrum, can be used to assign vibrational and rotational temperatures to the nitrogen molecule. Because the fluorescence spectrum of this electronic transition is very well understood and documented, this method allows
the determination of the internal energy of the ground electronic state molecule
to be obtained from the excited state fluorescence. This depends on the fact that
the fast electron ionization and excitation process is a vertical excitation, which
does not alter the ground electronic state atomic motion.[6]

Detailed analysis of the fluorescence spectra also requires a knowledge of the
Franck–Condon factors for the vibrational transitions, and the transition probabilities for the rotational branches. Since these quantities are well known in the
case of nitrogen,[7] and because the excitation and emission cross sections are

high, EBIF is a sensitive, useful method for the study of the internal energy state of nitrogen.[8,9] Other diatomic molecules, such as oxygen, hydrogen, and carbon monoxide, have been detected by EBIF by collection of the total fluorescence to measure the density of the molecular species.[4] However, not much work has been done using this method to extract internal state information on these molecules because of the complexity of the electron excited fluorescence spectrum in the case of hydrogen and carbon monoxide, or because of the lack of accurate transition probabilities for the appropriate lines in the oxygen fluorescence spectrum.

Internal state selective information can be more generally obtained using the method of resonantly enhanced multiphoton ionization or REMPI.[5] Since the probability for an electronic excitation or ionization process is much higher than direct absorption in the ground electronic state vibrational manifold of most molecules, this approach provides a sensitive route to internal state information in the spectroscopic fine structure of the electronic excitation and subsequent ionization of the molecules of interest. In general, the electronically excited state of the molecule should be stable to dissociation, and individual rovibrational lines should be identifiable in the excitation spectrum. Once in the excited state, the molecule absorbs additional photons from the excitation source and is ionized. The ions are then readily collected by an appropriate extraction field, and an ion signal proportional to the population of the specific irradiated rovibrational line is recorded. REMPI has the advantage that essentially all of the ions formed in the multiphoton process can be collected, increasing the sensitivity of the method over laser-induced fluorescence,[10] where the photons emitted into all space are not efficiently detected. However, absolute rovibrational populations are difficult to determine using REMPI, since the state-specific ionization cross sections are not known accurately for most molecules. Thus, REMPI detection schemes generally rely on the use of calibration methods to determine populations. An effusive beam source at a particular temperature or a static gas sample at a particular temperature may have to be probed with the REMPI detection method to establish the ionization cross sections for specific rovibrational levels. Once the ions are formed state specifically, TOF methods can be used to extract translational energy information in this method as well.[11]

Measurement of the ground state populations of desorbing molecules using REMPI involves measuring the ion current as a function of the excitation wavelength, generally in the UV or VUV region of the spectrum, depending on the molecular species being studied. For laser fluences below saturation of individual transitions, the ion current is given by[12]

$$I_{ion} = C_{ion}\frac{W_{ion}}{W_{ion} + \alpha} I_{abs} \qquad (7.1)$$

where W_{ion} is the ionization rate and α is the spontaneous fluorescence rate, which acts as an ion loss mechanism. C_{ion} is the ion detection efficiency, which is a function of the extraction geometry and experimental design. I_{abs} is the

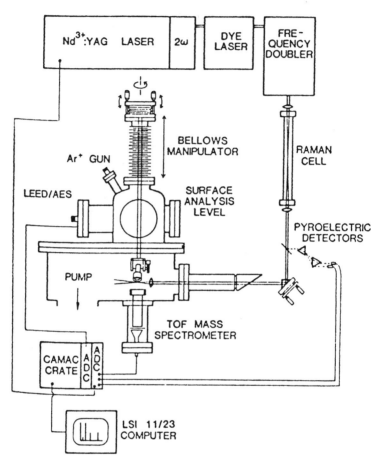

Figure 7.1.   Diagram of the overall experimental apparatus used to determine rovibrational state distributions of molecular hydrogen desorbing from clean copper surfaces. After ref. 27.

number of photons absorbed from the incident laser beam, and W_{ion} depends on the ionization cross section σ_{ion} and the ionizing laser intensity I_2. The cross section in general depends on the rovibrational state as well as the wavelength of the ionizing photon. For a single vibronic band, the measured ion signal is then directly proportional to I_{abs}, the number of photons absorbed from the excitation laser, as long as the ionizing laser intensity is constant and the fluorescence decay rate does not depend on the particular rotational state being probed. This is the case for a number of interesting molecular species, making REMPI a useful internal state selective detector for molecules such as NO, CO, N_2, OH, CH, NH_3, and SO_2. Some molecules, such as H_2, have ionization cross sections that depend strongly on the rovibronic state and the wavelength of the ionizing radiation,[13] making interpretation of populations a bit more complicated. In these cases it is essential to measure internal state populations of equilibrium samples of known temperatures to calibrate the measurements of an unknown distribution. A number of molecules have been examined using the REMPI method, and new REMPI schemes are being developed and applied to an increasing number of molecules of interest in heterogeneous reaction dynamics. Figure 7.1 shows a schematic diagram of an experimental apparatus that has been used for REMPI detection of the internal state distribution of molecular hydrogen desorbing from a copper single crystal surface.[14] This system forms one of the three examples discussed below.

7.3 Examples

7.3.1 Nitrogen Atom Recombination on Iron

The first heterogeneous system to be examined for internal energy excitation in the product of the surface reaction was the recombination of nitrogen atoms on the polycrystalline iron surface.[15] Previous work on the dynamics of heterogeneous atom recombination had concentrated on the measurement of the angular distribution, and, in favorable instances, the velocity distribution of the desorbing molecular product. The direct measurement of the internal energy state of a desorbing product molecule adds another dimension to the dynamic description of the heterogeneous process, and provides the sort of detailed information that is essential to the development of an understanding of the potential energy surface that governs heterogeneous reactions. In this early measurement, the surprising observation was made that the desorbing nitrogen molecules are significantly vibrationally excited.

The measurement was carried out using electron beam-induced fluorescence, EBIF,[4] which has been described briefly above. The apparatus used for these measurements is illustrated in Figure 7.2. As with earlier studies of hydrogen atom recombination on palladium and copper surfaces, the adsorbed atoms are supplied to the surface by permeation through a thin membrane. In the present case, the membrane is machined from a high purity polycrystalline iron rod, and welded to the end of a stainless steel tube that is fitted into a small furnace for heating the sample.[6] The membrane is polished after mounting on the stainless steel tube, to produce a membrane thickness of about 0.25 mm. Great care must be taken to avoid leaks in the membrane or in the weld attaching it to the supply tube. Molecular nitrogen is then introduced into this permeation source tube at a pressure slightly above atmospheric, and the molecules adsorb dissociatively on the hot back surface of the membrane and permeate the iron atomically at a rate determined by the temperature of the iron, the nitrogen pressure, the membrane thickness, and the activation energy for diffusion of nitrogen in iron.[16] These permeating atoms then recombine on the vacuum side of the membrane and desorb into the vacuum system where they interact with the electron beam as indicated in Figure 7.2.

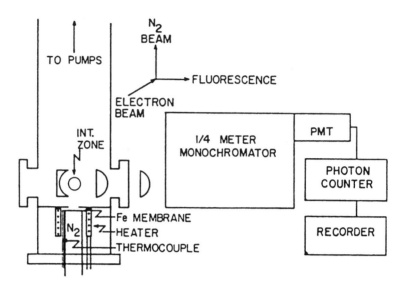

Figure 7.2.   Schematic diagram of experimental system used for N atom recombination studies. N_2 desorbing from the iron membrane is excited by an electron beam in the interaction zone. The fluorescence emission is focused on the entrance slit of the monochromator. After ref. 6.

Here the nitrogen molecules are ionized and excited to the N_2^+ $B^2\Sigma_u^+$ state by collisions with the electron beam. Fluorescence from this state is collected and focused on a monochromator, where the ratio of the vibrational band intensities can be used to assign a vibrational temperature to the desorbing nitrogen molecules. Observed vibrational band intensities following nitrogen molecule desorption from a membrane at 1150 K are shown in Figure 7.3. The 0-0, 0-1, and 1-0 bands for nitrogen emission are seen in the spectrum, along with two weaker CO+ related bands due to CO present in the vacuum system. The vibrational temperature obtained from these intensities is found to be around 2500 K, well above the temperature of the iron membrane. As the reaction was continued, the measured vibrational temperature was found to decrease. This observation was found to be related to the presence of significant amounts of sulfur on the membrane initially, with the coverage of sulfur decreasing as the membrane was heated for longer times. These observations suggest that the nitrogen molecule does not stay on the surface very long following a concerted recombination process on the iron surface.

Further studies of this system concentrated on the determination of the rotational state distribution of the desorbing nitrogen molecules, in addition to their vibrational energy.[17] They also addressed more specifically the effects of sulfur contamination on the observed internal energy excitation of the product species. Desorption of nitrogen molecules from a clean polycrystalline iron surface was found to result in nitrogen molecules with vibrational temperatures only slightly

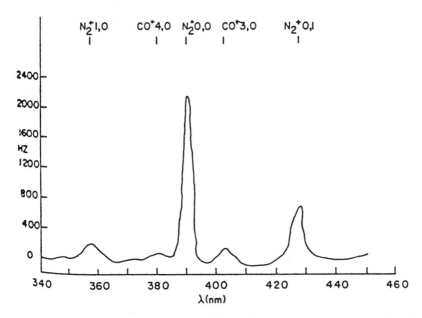

Figure 7.3. Fluorescence spectrum from 350 to 430 nm after background subtraction. Intensities used for temperature determination are corrected for overlapping CO contributions. After ref. 15.

greater than the temperature of the surface. The composition of the iron surface was monitored before and after the desorption measurements using an in situ soft X-ray appearance potential spectrometer.[6] Over the surface temperature range from 1086 to 1390 K, the measured vibrational temperature was found to be essentially constant at around 1300 K. For these same clean surfaces, and over the same surface temperature range, the rotational temperature measured for the desorbing nitrogen molecules was found to be 400 ± 30 K, irrespective of surface temperature.

When these measurements were carried out on iron membranes contaminated by sulfur, the measured vibrational temperature was found to be significantly higher than the temperature of the surface. Over the sulfur coverage range from about 0.1 to 0.3 monolayer, the measured vibrational temperature was found to increase from 1600 to about 2600 K, for a surface temperature of 1150 K, in contrast to what was observed on the clean iron surface. Over this same range of sulfur coverages, the measured rotational temperature was constant at 400 K, as was observed for the clean iron membranes.

These observations imply particular details about the interaction potential energy surface governing the nitrogen atom recombination on iron. The low rotational temperature observed, essentially independent of surface temperature or composition, suggests that there is a steric constraint for those adsorbed atoms that recombine and desorb on the iron surface. Recombination events are likely to be "head-on," small impact parameter events on the iron surface, with the small resulting tangential component of the angular momentum of the collision of the atoms on the surface being reflected in the low measured rotational temperature. Sulfur coverage would not be expected to alter this steric constraint, if the recombination is envisioned as occurring in the channel regions defined by the iron atoms on the low index bcc surfaces. Sulfur would effectively block these channel sites, reducing the amount of desorbing molecular nitrogen, as was observed, but not significantly affecting the steric constraint and the resulting measured rotational temperature.

The measured vibrational temperature of the desorbing nitrogen molecules and its dependence on sulfur coverage suggest that these observations probe the nitrogen–iron interaction potential normal to the surface. In particular, these results suggest that the vibrational temperature is determined by the height and position of the activation barrier for dissociative adsorption of nitrogen on iron. This can be best understood by reference to gas phase potential energy surfaces for atom–diatom collisions resulting in atom exchange.[18] This is seen in Figure 7.4, which illustrates the effect of the position of the activation energy crest along the reaction coordinate for the atom exchange reaction. An early barrier results in product vibrational excitation, while decreasing vibrational excitation is observed as the barrier moves toward the product end of the reaction coordinate. In this work, it was postulated that the coverage of sulfur on the surface served to shift the position of the activation barrier closer to the reactant end of the reaction coordinate, resulting in vibrational excitation of the product. This shift in activation barrier occurs when the atomic nitrogen binding energy is increased by the presence of sulfur on the surface. Other electronegative adsorbates should also affect the internal energy of the product nitrogen molecules, allowing a further test of this proposed model. Similarly, other structural arrangements of substrate atoms (specific low index surfaces) could be used to test the steric constraint model proposed to explain the rotational temperature observations.

7.3.2 Hydrogen Atom Recombination on Copper

Perhaps the most extensively studied heterogeneously catalyzed atom recombination reaction is the copper-catalyzed recombination of hydrogen atoms. This reaction and its inverse, the dissociative adsorption of hydrogen on copper, has been studied in detail by several groups over the past 20 years. Molecular hydrogen adsorption on copper surfaces has long been known to be an activated process. The activation energy for dissociative adsorption has been estimated to be on the order of 8–10 kcal/mol, while the heat of adsorption of hydrogen on copper is in the range of 10–12 kcal/mol.[19] Detailed thermal desorption studies of the adsorption of hydrogen on Cu(110) indicate that the second-order preexponential factor for the recombination and desorption of D atoms from this surface is about 10^4 smaller than for most second-order desorption processes.[20]

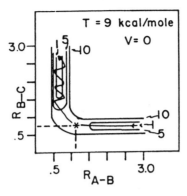

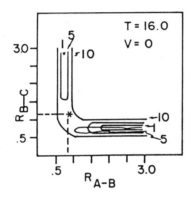

Figure 7.4. Potential energy surfaces for the atom–diatom exchange reaction: A + B − C → A − B + C. Note the position of the activation barrier, denoted by an asterisk (*), with respect to the reaction coordinate. After ref. 17.

This low value is suggestive of a rather constrained desorption transition state, and is indicative of unusual desorption dynamics for this system.

Initial direct measurements of the desorption dynamics were carried out in the mid 1970s by Stickney.[21,22] This group monitored the angular distribution of hydrogen flux following the recombinative desorption of hydrogen on the Cu(111), Cu(110), and Cu(100) surfaces, and found that the angular distributions of the desorbing molecules were very highly peaked at the surface normal. Rather than the $\cos(\theta)$ behavior expected for equilibrium desorption characterized by the temperature of the surface, Balooch and Stickney[22] observed angular distributions that were fit to the form $\cos^n(\theta)$, where n ranged from 2.5 for the (110), to 5 for the (100), and 6 for the (111) surfaces. This behavior was ascribed to the possible presence of an activation barrier to dissociation resulting from the crossing of the curves for atomic and molecular adsorption on the surface. The presence of such an activation barrier for dissociative adsorption inferred from the angular distribution of recombinatively desorbing molecules should be reflected in an angular and translational energy-dependent behavior directly in the dissociative adsorption process.

This was actually observed in the later measurements of Balooch and co-workers,[23] where the dissociative adsorption probability of hydrogen on copper single crystal surfaces was measured carefully as a function of incident angle and translational energy. They observed a threshold to dissociative adsorption as energy and incident angle were varied, which could be described as a one-dimensional (normal) activation barrier for the dissociative adsorption.[24] Barrier heights that depended on the specific crystal face were derived, which were in the range of 3–5 kcal/mol. Using the principle of detailed balance, these angle and velocity-dependent adsorption results were used to calculate the velocity-integrated desorption angular distributions measured earlier by this group.[25] When TOF methods were used to measure the desorbing molecule velocities as a function of desorption angle, however, the correlation between desorbing angle and velocity predicted by the simple one-dimensional barrier to desorption due to Van Willigen,[24] which was applied successfully to the velocity-integrated angular distributions,[25] was not observed.[26] The one-dimensional barrier model predicts that the measured velocity should increase with increasing angle from the surface normal, and this was not observed. Rather, sharp velocity distributions with mean translational energies greater than 3500 K were observed for the (100) and (111) surfaces. These results suggest that the potential energy surface controlling postpermeation recombinative desorption may not be the same as the potential energy surface governing dissociative adsorption on copper. More will be said about this possibility when we discuss internal energy measurements for this system, and when we discuss the desorption of hydrogen from the palladium surface, the third example of this chapter.

In a series of very elegant experiments, Kubiak et al.[27] examined the internal energy of hydrogen and deuterium desorbing from Cu(110) and (111) single crystal surfaces. The permeation recombination approach was again used in these measurements, providing a steady flux of desorbing hydrogen. The product molecules were probed using a resonantly enhanced multiphoton ionization (REMPI) $(2 + 1)$ detection scheme.[28] Two photon excitation of hydrogen to the E, $F^1\Sigma_g^+$ state was accomplished using 193 and 211 nm radiation from a hydrogen Raman shift cell pumped by the doubled output from a pulsed dye laser pumped by a frequency doubled YAG laser. The doubled output of the dye laser is tunable in the region near 285 nm, and is focused into the hydrogen Raman shift cell. The third and fourth anti-Stokes lines at 211 and 193 nm are used for the excitation and ionization of the desorbing hydrogen or deuterium molecules. Measured rotational distributions are found to be slightly below the temperature of the copper surface,[29] and are identical within experimental error for hydrogen and deuterium. Additionally, the data for the two different surface orientations are identical, and no systematic differences in the rotational state distribution can be seen for hydrogen in the $v'' = 1$ or $v'' = 0$ vibrational bands. The distributions are slightly non-Boltzmann, although not markedly so, and there are no apparent differences between *ortho* and *para* hydrogen desorbing from the surfaces.

In contrast to the below surface temperature rotational state distributions, REMPI studies of the vibrational populations show hydrogen molecules that are dramatically vibrationally excited above equilibration at the surface temperature. The ratio of population in the $v'' = 1$ to $v'' = 0$ states suggests an excess of population in the first vibrationally excited state 60–90 times higher than would be expected from equilibrium desorption at the temperature of the surface. The excitation is somewhat greater on the Cu(111) surface than on the Cu(110) surface. There is no significant population observed in the $v'' = 2$ band, suggesting that the vibrational excitation is not well characterized by a high vibrational temperature, but that the ratio of first excited state population to ground state population is clearly hot but nonequilibrium.

The rotational state distribution observed in these measurements, and its independence of surface crystallographic orientation, surface temperature, and vibrational mode, suggests that the recombination occurs in a region of the repulsive potential energy surface that is not highly structured. Repulsive potentials that could give rise to angular momentum to the desorbing molecule appear to be quite flat in the region where the recombination occurs. Kubiak et al.[27] estimate that recombination occurs about 3.75 Å above the (110) trough atoms of the copper surface.[27] This is an alternative explanation for the low rotational temperatures observed in the N atom recombination case described above, where low rotational energy was ascribed to small impact parameter collisions of the recombining atoms in the highly structured trough regions of the substrate.[17] For the case of H atom recombination on the copper surface, recombination in the trough would form molecules in a region of the potential energy surface corresponding to repulsive energies of several electron volts. This is considerably higher than the low rotational energy observed, and suggests that the recombination must be occurring in regions of the potential surface with less electron density than that in the (110) troughs on this surface.

Since the desorbing hydrogen molecules are found to be vibrationally excited, with the ratio of the first excited state to the ground state population a factor of 50 to 100 greater than expected for equilibration at the surface temperature, detailed balance arguments suggest that the adsorption probability should be enhanced by similar factors for molecules incident on the surface in the first vibrationally excited state. Kubiak et al.[27] attempted to use detailed balance arguments to calculate the excess vibrational excitation in desorbing hydrogen, using the adsorption measurements of Balooch and co-workers[23] described earlier. In this case, incident hydrogen and deuterium beam source temperatures were used to estimate the degree of vibrational excitation present in the incident source beams of the work by Balooch et al.,[23] and the data on dissociative adsorption as a function of incident energy, angle, and vibrational excitation was used to calculate the expected vibrational excitation in hydrogen desorbing from the copper surface. These calculated values were then compared to the measurements of Kubiak et al.[27] This comparison is seen in Figure 7.5, and it is evident that the detailed balance argument does not work here. This suggests again that the postpermeation recombinative desorption is likely to occur on a different potential energy surface than that governing dissociative adsorption of molecules incident on the surface from the gas phase.

Figure 7.6 suggests a possible difference in the potential energy surfaces, which may explain the observed dynamical differences between postpermeation recombinative desorption and dissociative adsorption on the copper surface. The energetics in this figure indicate that a dissolved H atom diffusing to the copper surface from the bulk may have as much as 21 kcal/mol more energy than the

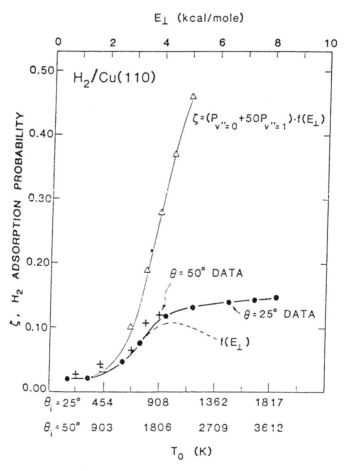

Figure 7.5. Comparison of the H_2/Cu(110) adsorption probability data and the detailed balance prediction. The prediction uses the result that the $P_{v''=1}/P_{v''=0}$ ratio desorption is ~50 times greater than the equilibrium at T_s value. After ref. 27.

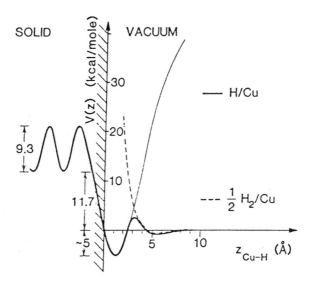

Figure 7.6. Schematic one-dimensional potential energy curves for hydrogen on copper. After ref. 27.

hydrogen molecule in the gas phase. When an energetic absorbed atom reacts with an adsorbed atom at the surface before the absorbed atom is equilibrated in the chemisorption well, a significant amount of reaction energy must be lost to the surface prior to desorption at equilibrium. This relatively inefficient process could then result in desorbing hydrogen molecules that are vibrationally and translationally excited. This energetic argument is also consistent with the fact that no molecules excited into the $v'' = 2$ vibrational level are detected. The energy of this level, at 23.2 kcal/mol above the ground state, is greater than the energy available from the unrelaxed absorbed atom. Excitation into this level prior to desorption would require an unlikely recombination event involving two unrelaxed absorbed atoms forming the desorbing molecule.

The detailed mechanism of the recombination process that results in the surface normal directed angular distribution, the translational excitation, and the vibrational excitation of the desorbing molecules has been addressed in some detail in two interesting theoretical studies. A classical trajectory calculation by Harris et al.[30] used a potential energy surface derived from electronic structure calculations for the adsorption of hydrogen on copper. A main feature of the potential energy surface is the presence of an entrance channel activation barrier. The saddle point in the potential energy surface, which separates desorbing molecules from the chemisorbed atoms, corresponds to a lengthened H–H bond relative to its value in the gas phase. This suggests that a portion of the activation energy for the recombination reaction will appear as vibrational excitation in the product, in addition to the translational excitation and focusing of the desorbing molecules in the direction normal to the surface. This potential energy surface is illustrated in Figure 7.7. The dashed line on this figure indicates the boundary or seam of the potential energy surface separating the chemisorption well from the molecular exit channel. The cross indicates the saddle point on the PES, where the force on the system is zero. All along the seam, the force on the particles is directed toward this saddle point.

A series of classical trajectories was run on this potential surface, with initial conditions chosen so that equilibrated hydrogen atoms located at points along the seam with total energies ranging from zero to a few kT_s would interact to cross the seam and desorb from the surface through the exit channel. This approach of restricting trajectories to initial conditions near the seam in the PES is useful for studying the rare recombination event. A random choice of initial conditions would have the trajectories dominated by atoms deep in the chemisorption well interacting but never escaping the surface. This sort of half-collision approach is quite common in gas phase trajectory calculations,[31] and is certainly appropriate for the study described here. It was further assumed that the probability that a desorbing molecule crosses the seam with excess kinetic energy ϵ is given by a Boltzmann distribution.

$$P_s \sim \exp\left[-\beta(E - E^* + \epsilon)\right] \tag{7.2}$$

where P_s the probability of crossing at a point s on the seam, E_s is the system energy at that point, and E^* is the saddle point energy.

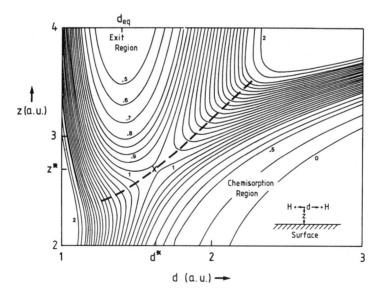

Figure 7.7. Energy contours used in trajectory calculations. The seam marking the boundary between the chemisorption region and the exit channel is marked with a dashed line. The cross denotes the saddle point where the force on the system is zero. After ref. 30.

A large number of trajectories was then calculated, and the translational, vibrational, and rotational energies of the molecules traveling out the exit channel were weighted with this Boltzmann factor and tabulated. It was found that the desorbing molecule trajectories exhibited significant translational excitation and forward focusing, although sufficient trajectories were not run to provide accurate estimates of the actual angular distribution. A large fraction of the desorbing molecules were vibrationally excited, due to the extension of the H–H bond at the position of the saddle point, and an estimate of the first excited state population to the ground state vibrational population was made. This estimate assumes that the quantum state distributions correspond to the average energy of the classical trajectory calculated distributions. Using this approach, the P_1/P_0 ratio was estimated as 0.25, compared to the experimentally measured value of 0.08 determined by Kubiak et al.[27] The rotational energy of the desorbing molecules was found to be low, and to decrease with surface temperature.

The other theoretical treatment of this experimental system is the negative ion potential model proposed by Gadzuk.[32] As described by Kubiak et al.[27] a half-collision application of the hydrogen scattering model proposed by Gadzuk might be applicable to the vibrational excitation results of their work. Adsorbed H atoms with partial negative charge form a molecule on a H_2^-/Cu potential energy surface. The newly formed molecule will travel on this PES as it moves away from the surface. At some point above the surface, the negative charge transfers to unoccupied states in the copper substrate. Depending on the total energy of the two atoms that made this molecule, and the distance at which the electron transfer takes place, the molecule may leave the surface vibrationally excited. If the electron transfer is sudden, the probability of formation of a molecule in the first excited state relative to the ground state will depend on the Franck–Condon factors that connect the negative ion and neutral potential energy surfaces.[33] This ion-neutral curve crossing mechanism can explain the observation of vibrationally excited desorbing molecules, and could be further tested by examining the degree of hydrogen molecule vibrational excitation as a function of the work function of the surface, which should control the extent of charge transfer between the negative ion state and the surface. Kubiak et al.[27] noted a sharp decrease in the observed vibrational excitation for the sulfur-modified copper surface,[27] whose work function increases by about 0.30 eV, with sulfur coverage consistent with this picture.[34]

7.3.3 Hydrogen Atom Recombination on Palladium

The third example we discuss is another system that has been extensively studied. The postpermeation recombination and desorption of hydrogen from various palladium surfaces have been investigated carefully over the past 15 years. Experimentally, the hydrogen–palladium system is convenient because of the high permeability of hydrogen in palladium,[35] making the supply of hydrogen to the surface relatively facile, and the subsequent desorption signal significant over a wide surface temperature range. This system provided the first direct TOF measurement of the velocity of the recombination product molecules,[36] using a psuedo-random chopper method[37] to unambiguously observe a bimodal velocity distribution for deuterium desorbing from a Pd(100) surface.

The results of this initial measurement are shown in Figure 7.8. The TOF spectra are shown for several detection angles for a Pd(100) surface covered with about one-half monolayer of sulfur. The bimodal nature of the velocity distribution is clearly seen at detection angles of 0 and 20° with respect to the surface normal. At higher angles, the broader, slower velocity distribution dominates. The narrow, fast distribution corresponds to an angular dependence of nearly $\cos^{10}(\theta)$, while the slower (Maxwellian) section of the distribution has an angular dependence closer to $\cos(\theta)$. The presence of sulfur on the surface clearly controls the measured velocity distributions. For the clean Pd(100) surface, only the slower, broader velocity distribution molecules are observed at all detection angles. As the sulfur coverage increases, the fast molecules desorbing normal to the surface become evident. The overall behavior appears to be a superposition

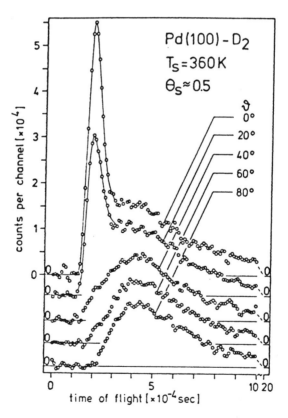

Figure 7.8. Sequence of TOF curves with desorption angle θ as parameter. After ref. 36.

of two distinct velocity distributions, one characteristic of the clean surface and the other of the surface blocked by sulfur or other impurities.

Some time later, Zacharias and co-workers probed the internal energy of the hydrogen desorbing after atom recombination on polycrystalline palladium, using laser-induced fluorescence.[38] In these experiments, the polycrystalline foil was not compositionally characterized in situ, but was cleaned by heating to 700 K for several hours prior to making desorption measurements. Vacuum UV laser–induced fluorescence was used to single photon excite the $B^1\Sigma_u^+$ ($v'' = 3$) $\leftarrow X^1\Sigma_g^+$ ($v'' = 0$) transition near wavelengths of 106 nm. Rotationally resolved excitation spectra were obtained, which allowed the determination of the rotational temperature of the desorbing hydrogen at two surface temperatures. The measured rotational temperature was slightly below the surface temperature in each case, and there was evidence for a non-Boltzmann distribution for the low J states.

This group followed up these initial measurements with a more detailed study of the internal energy of deuterium desorbing from the clean Pd(100) surface.[39] Auger spectroscopy and low-energy electron diffraction were used to characterize the Pd(100) sample, and the permeation geometry of previous studies was used to supply the deuterium atoms to the surface of the palladium sample. A (1 + 1) REMPI scheme was used to state selectively detect the desorbing deuterium molecules. The first photon at 106 nm excites the molecule in the $B^1\Sigma_u^+ \leftarrow X^1\Sigma_g^+$ system, and a second photon at 318 nm from the initial pump laser is used to selectively ionize the excited molecules. Using this approach, the vibrational population in the first excited state was determined to be 1.5 ± 0.3% for desorption at a surface temperature of 677 K. This is about a factor of 10 greater than would be expected for desorption at equilibrium at this surface temperature. This population is somewhat lower than that observed for hydrogen or deuterium recombination desorption on clean copper surfaces.

Time of flight measurements of the vibrationally excited molecules indicate that the velocity of these molecules desorbing from the clean Pd(100) surface is the same as the velocity observed for the ground vibrational state molecules. Within experimental error, the velocity of both states is Maxwellian at the temperature of the surface. The observation of Arrhenius like behavior for the intensity of the vibrationally excited species as a function of surface temperature suggest that the mechanism for vibrational excitation in this case is purely thermal. Rotational temperature measurements for deuterium and hydrogen desorbing from the clean Pd(100) surface indicated a rotational temperature always less than the surface temperature over the range of surface temperature from 300 to about 800 K.[40] This observation suggests the existence of a molecular precursor state in hydrogen desorption from this surface, a suggestion that is consistent with calculations for hydrogen adsorption on platinum,[41] and for spectroscopic observations of hydrogen adsorption on copper and silver surfaces.[42]

Rovibrational state-specific TOF measurements have also been made on the desorbing deuterium molecules. Figure 7.9 shows a typical TOF spectrum for hydrogen molecules desorbing from the clean Pd(100) surface in the ($v = 0$,

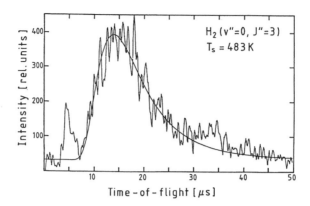

Figure 7.9.   State-specific TOF spectrum of $H_2(v'' = 0, J'' = 3)$ desorbing from Pd(100).
After ref. 43.

$J = 3$) state at a surface temperature of 485 K. The short time peak is due to
atomic hydrogen ions produced by multiphoton dissociation and ionization. The
broader peak corresponds to the hydrogen molecular ion, and the line is a fit to
a Maxwellian flux distribution at 450 K. This is slightly below the surface temperature, but within the experimental uncertainty. Similar behavior is observed
for the other rovibrational TOF spectra measured for this system, including the
rotational lines of the first excited vibrational level.[43] This observation suggests
a decoupling of the translational, rotational, and vibrational degrees of freedom
in the desorbing molecule.

7.4 Comments

This chapter has shown the degree of dynamic detail that is now available in
studies of simple heterogeneous reaction systems. The use of laser-induced fluorescence, REMPI, and related methods allows state-specific probing of the
products of surface reactions. The concentration on recombination desorption
reactions, particularly hydrogen, has provided a wealth of information that is
stimulating a good deal of theoretical activity. As with the previous chapters,
the examples presented here also suggest areas that are ripe for further work.

   In each of the examples in this chapter, it is clear that a detailed knowledge
of the surface structure and composition of the substrate is crucial to the complete
understanding of the reaction dynamics. The composition of the surface affects
the dynamics, so it must be known and controllable for experiments of this sort
to make sense. More importantly, models for the processes studied here can be
tested by control of the surface composition, as is seen in the example of the
effect of work function on the negative ion PES model for hydrogen desorption
from copper. Similarly, careful control of the crystallographic structure of the

substrate can allow the testing of geometric constraint models, as suggested in the recombinative desorption of nitrogen from iron. It is also clear that very low levels of contaminant or structural defects such as steps can have a controlling influence on the observed dynamics. It is imperative that as more laser and spectroscopic schemes are developed to probe the internal energy or velocity distribution of more molecules of interest to heterogeneous dynamics, that researchers not lose sight of this fact. Careful surface science and in situ characterization of the substrate must be coupled with clever detection schemes and spectroscopic analysis.

It is also clear from the examples presented here that the more detailed the dynamic information that can be obtained, the clearer the understanding that can be developed. Rovibrationally state-specific velocity distributions as a function of desorption angle comprise the ideal data set, which obviously is very difficult and tedious to obtain. This data for a range of surface parameters (temperature, composition, structure) then allows for the development of models that can be rigorously tested. Polarization-dependent measurements could also be carried out, which would provide information about the orientation of the angular momentum vector for rotating desorbed molecules. The inclusion of results from the sort of scattering measurements described in Chapters 3 and 6 will also help develop these theoretical descriptions.

Theoretical input is also needed for the development of understanding in this area. Classical trajectory calculations were mentioned in the examples discussed here, and are likely to continue to contribute to the physical understanding of the dynamics of these simple surface reactions. Trajectory calculations rely on the availability of accurate potential energy surfaces, and much work is needed to provide reliable ab initio methods for calculating these complex many body systems. Quantum mechanical modeling, quantum scattering calculations, and work on data inversion algorithms are all areas where progress would stimulate further detailed experimentation, which in turn would stimulate more theoretical work. The interaction of theory and experiment is essential for progress in understanding to occur in these studies, and the present level of development of theoretical and experimental tools promises a rapid increase in our developing understanding.

References

1. G. Comsa and R. David, *Surface Sci. Rep.* **5**, 145 (1985).

2. H. Zacharias, *Int. J. Mod. Phys. B* **4**, 45 (1990).

3. P. L. Houston and R. P. Merrill, *Chem. Rev.* **88**, 657 (1988).

   J. A. Barker and D. J. Auerbach, *Surface Sci. Rep.* **4**, 1 (1985).

   M. C. Lin and G. Ertl, *Annu. Rev. Phys. Chem.* **37**, 587 (1986).

4. E. P. Muntz, *The Electron Beam Fluorescence Technique* AGARDograph 132, 1968.

5. P. M. Johnson, *Acc. Chem. Res.* **13**, 1 (1980).

6. R. P. Thorman and S. L. Bernasek, *Rev. Sci. Instrum.* **52**, 553 (1981).

7. G. Herzberg, *Molecular Spectra and Molecular Structure, I. Spectra of Diatomic Molecules,* Van Nostrand, New York, 1950.

8. E. P. Muntz, *Phys. Fluids* **5**, 80 (1962).

9. P. B. Scott and T. Mincer, *Entropie* **30**, 170 (1969).

10. R. N. Zare and P. J. Dagdigian, *Science* **185**, 739 (1985).

11. H. Zacharias, *Appl. Phys* **A47**, 37 (1988).

12. H. Rottke and H. Zacharias, *J. Chem. Phys.* **83**, 4831 (1985).

13. D. C. Jacobs and R. N. Zare, *J. Chem. Phys.* **85**, 5457 (1986); D. C. Jacobs, R. J. Madix, and R. N. Zare, *J. Chem. Phys.* **85**, 5469 (1986).

14. G. D. Kubiak, G. O. Sitz, and R. N. Zare, *J. Vac. Sci. Technol.* **A3**, 1649 (1985).

15. R. P. Thorman, D. Anderson, and S. L. Bernasek, *Phys. Rev. Lett.* **44**, 743 (1980).

16. R. M. Barrer, *Diffusion in and Through Solids,* Cambridge University Press, Cambridge, England, 1941.

17. R. P. Thorman and S. L. Bernasek, *J. Chem. Phys.* **74**, 6498 (1981).

18. J. C. Polanyi, *Acc. Chem. Res.* **5**, 161 (1972).

19. J. Pritchard and F. C. Tompkins, *Trans. Faraday Soc.* **56**, 540 (1960).

    C. S. Alexander and J. Pritchard, *J. Chem. Soc. Faraday Trans.* **68**, 202 (1972). J. Pritchard, T. Catterick, and R. K. Gupta, *Surface Sci.* **53**, 1 (1975).

20. I. E. Wachs and R. J. Madix, *Surface Sci.* **84**, 375 (1979).

21. T. L. Bradley and R. E. Stickney, *Surface Sci.* **38**, 313 (1973).

22. M. Balooch and R. E. Stickney, *Surface Sci.* **44**, 310 (1974).

23. M. Balooch, M. J. Cardillo, D. R. Miller, and R. E. Stickney, *Surface Sci.* **46**, 358 (1974).

24. W. van Willigen, *Phys. Lett.* **A28**, 80 (1968).

25. M. J. Cardillo, M. Balooch, and R. E. Stickney, *Surface Sci.* **50**, 263 (1975).

26. G. Comsa and R. David, *Surface Sci.* **117**, 77 (1982).

27. G. D. Kubiak, G. O. Sitz, and R. N. Zare, *J. Chem. Phys.* **83**, 2538 (1985).

28. (a) E. E. Marinero, C. T. Rettner, and R. N. Zare, *Phys. Rev. Lett.* **48**, 1323 (1982); (b) E. E. Marinero, R. Vasudev, and R. N. Zare, *J. Chem. Phys.* **78**, 692 (1983); (c) S. L. Anderson, G. D. Kubiak, and R. N. Zare, *Chem. Phys. Lett.* **105**, 22 (1984).

29. G. D. Kubiak, G. O. Sitz, and R. N. Zare, *J. Chem. Phys.* **81**, 6397 (1984).

30. J. Harris, T. Rahman, and K. Yang, *Surface Sci.* **198**, L312 (1988).

31. R. A. LaBudde, P. J. Kurty, R. Bernstein, and R. D. Levine, *J. Chem. Phys.* **59**, 6286 (1973).

32. (a) J. W. Gadzuk, *J. Chem. Phys.* **79**, 6341 (1983); (b) J. W. Gadzuk and J. K. Nørskov, *J. Chem. Phys.* **81**, 2828 (1984).

33. (a) J. W. Gadzuk, *J. Chem. Phys.* **79**, 3982 (1983); (b) *Phys. Rev. B* **20**, 515 (1979).

34. G. G. Tibbetts, J. M. Burkstrand, and J. C. Tracy, *Phys. Rev. B* **15**, 3652 (1977).

35. W. van Willigen *Phys. Letters* **28A**, 80 (1968)

36. G. Comsa, R. David, and B-J Schumacher, *Surface Sci.* **95**, L210 (1980).

37. F. Gompf, W. Reichardt, W. Gläser, and K. H. Beckurts, in *Neutron Inelastic Scattering,* Vol. II, IAEA, Vienna, 1968, p. 417.

38. H. Zacharias, *Chem. Phys. Lett.* **115**, 205 (1985).

39. L. Schröter and H. Zacharias, *Phys. Rev. Lett.* **62**, 571 (1989).

40. L. Schröter, G. Ahlers, H. Zacharias, and R. David., *J. Electr. Spectr. Rel. Phen.* **45**, 403 (1987).

41. J. E. Miller, *Phys. Rev. Lett.* **59**, 2943 (1987).

42. Ph. Avouris, D. Schmeisser, and J. E. Demuth, *Phys. Rev. Lett.* **48**, 199 (1982). S. Andersson and J. E. Harris, *Phys. Rev. Lett.* **48**, 545 (1982).

43. H. Zacharias, *Appl. Phys.* **A47**, 37 (1988).

8

Catalytic Oxidation

Perhaps the most extensively studied "complex" heterogeneous reaction is the catalytic oxidation of CO on the surface of platinum metals. Over the past decade, careful examinations of the kinetics of this reaction on well-characterized surfaces using molecular beam scattering methods have been performed. In addition, the catalytic oxidation of CO has been investigated using dynamics methods that probe the internal energy of the product CO_2 formed in this exothermic surface reaction. These investigations were the first to obtain internal state information about the product of a heterogeneous reaction more complex than the recombination of adsorbed atoms. A reasonably detailed understanding of this reaction has been obtained, and it serves as a good example of the sort of kinetic and dynamic information that can be developed about heterogeneous reactions using the methods of molecular reaction dynamics in combination with the tools of surface chemical physics.

8.1 Experimental Methods

In studies of the kinetics of the catalytic oxidation of CO on well-characterized surfaces, modulated molecular beam mass spectrometry[1] has figured very prominently. This method uses a beam of reactant molecules incident on the well-characterized surface in an ultrahigh vacuum scattering chamber to probe the reaction kinetics and mechanism at the surface. The reactant beam is mechanically chopped with a rotating chopper, which modulates the concentration of reactant on the surface. Desorbed product is monitored with a mass spectrometric

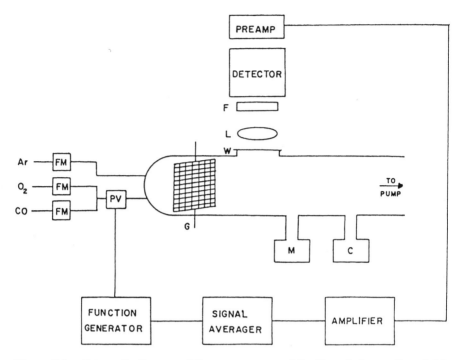

Figure 8.1.   Schematic diagram of flow apparatus used for IR emission studies of CO oxidation.

detector, and the waveform of the desorbed product signal is used to extract reaction mechanism and kinetic information. In the case of catalytic oxidation of CO, the incident-modulated reactant beam may be O_2, and the surface is exposed to a steady-state background pressure of CO. Scattered, unreacted O_2 as well as product CO_2 is monitored as a function of angle of the detector from the surface normal. This method has been discussed in some detail in Chapter 6, and a schematic diagram of an apparatus used for modulated molecular beam mass spectrometry was shown there.

Infrared emission techniques have been successfully applied to the study of the dynamics of the catalytic oxidation reaction. Work carried out simultaneously by Haller and co-workers[2] and by Bernasek and Leone[3] showed the power of this approach for obtaining dynamics information about this reaction. The catalytic oxidation of CO on the platinum group metals is particularly amenable to study using IR emission methods for two reasons. First, the oxidation reaction is exothermic overall, and might be expected to deposit some of that reaction exothermicity into the CO_2 product molecule. Second, the product CO_2 is very weakly bound to platinum, so that the likelihood for equilibration of the nascent excited CO_2 with the surface prior to desorption is reduced. This combination of conditions suggests that any internal state information obtained by emission (or adsorption) spectroscopy of the product CO_2 will be related to the dynamics of the surface process that forms the product molecule.

Figure 8.1 shows a simple schematic diagram of a flow apparatus that was used to monitor IR emission from the CO_2 product of the CO oxidation reaction on a polycrystalline platinum gauze. Various filtering methods can be used to examine the emission from the excited product molecule. Narrow and wide band interference filters provide an estimate of the total internal excitation of the product molecule. Cold gas filtering methods,[4] where the product emission passes through a gas cell containing room temperature CO_2 gas, are used to provide an estimate of excitation in higher lying vibrational states of the product molecule. A circular variable filter is used to resolve the IR emission, providing an estimate of the degree of vibrational excitation in the product molecule.

Haller and co-workers have used Fourier transform interferometry to examine the IR emission of the CO_2 product of this reaction, by replacing the filter/detector arrangement of Figure 8.1 with an interferometer. This provides a very high resolution emission spectrum, which can then be fitted to spectra calculated for particular postulated internal state distributions. This approach has been applied to the CO_2 produced when a free jet expansion of CO and O_2 was incident on a polycrystalline platinum foil. The method provides very high resolution internal state information, but requires rather high reactant fluxes and very careful alignment of the interferometer to extract reliable signal levels. Both of the approaches discussed here and described in more detail in the case studies that follow suffer from the fact that to obtain usable signals, reactant pressures and catalyst surface areas have to be rather high. This makes careful characterization of the structure and composition of the catalytic surface difficult. The methods are promising, however, and are likely to be applied in experimental arrangements where standard UHV characterization tools will be usable.

8.2 Illustrative Examples

The catalytic oxidation of CO by oxygen on platinum and rhodium surfaces provides both of the illustrative examples discussed in this chapter. The first example concentrates on detailed study of the mechanism and kinetics of this important reaction using modulated beam mass spectrometry. The second example builds on this kinetic and mechanistic base, and provides some of the most detailed dynamic information about a heterogeneous reaction presently available.

8.2.1 Modulated Beam Mass Spectrometric Study of the Catalytic Oxidation of CO by O_2 on Platinum and Rhodium

A detailed series of studies was carried out early in the 1980s by Campbell, Ertl, and co-workers on the interaction of CO, O_2, and the CO oxidation on the well-characterized Pt(111) surface.[5–7] Using the modulated molecular beam methods described above, in combination with thermal desorption and angular distribution measurements, a detailed picture of this heterogeneous reaction was obtained. Carbon monoxide was found to adsorb on the Pt(111) surface with a sticking coefficient of 0.84, independent of angle of incidence and surface temperature. Modulated beam measurements indicated a very high mobility for the CO on this surface, and a localization of the absorbed CO at energetically favorable defect and step sites on the surface. Desorption of CO occurs from these defect sites, and exhibits a desorption activation energy of about 35 kcal/mol.[5]

A much lower dissociative sticking probability was observed for O_2 interacting with the Pt(111) surface. The initial sticking coefficient for O_2 decreases from a value of 0.65 at 300 K to 0.025 at 600 K. The measured desorption activation energy varies with oxygen coverage from 51 to 42 kcal/mol at high oxygen coverage. Angular distribution measurements of the desorbing oxygen flux show a slight peaking in the normal direction, suggesting that some of the desorbing oxygen is not fully accommodated molecularly prior to desorption.[6] LEED studies of this adsorption system indicate that islands of atomic oxygen are formed on the Pt(111) surface following dissociative adsorption of the molecular species,[8] most probably by way of a spectroscopically characterized peroxo species.[9]

Building on these studies of the adsorption and desorption of the CO and O_2 reactant molecules, Campbell et al.[7] undertook a detailed study of the catalytic CO oxidation on the Pt(111) surface. A measure of the steady-state CO_2 production as a function of surface temperature for a constant O_2 beam incident on the surface held in a background pressure of CO gave a maximum rate of CO_2 production at a surface temperature of about 440 K. The reaction was found to be first order in O_2 pressure over all temperatures studied. At low CO pressures, the rate was first order in CO pressure, but at higher CO pressures it was zero order for high surface temperature, or displayed negative order kinetics at lower temperatures (CO adsorption at low surface temperatures inhibited the reaction rate).

A modulated beam of O_2 incident on the platinum surface exposed to a constant background of CO resulted in a modulated product CO_2 signal whose phase and amplitude were monitored for a range of surface temperatures and reactant fluxes. Analysis of these data, along with titration transient measurements, clearly indicate that the CO oxidation on the Pt(111) surface proceeds via a Langmuir–Hinshelwood mechanism. That is, the reaction occurs between adsorbed species, mobile molecular CO, and atomic oxygen in this case. Figure 8.2 shows a one-dimensional potential energy diagram for this reaction in the low coverage region. Overall, the reaction is strongly exothermic, but the surface reaction is only weakly so. Most of the exothermicity is expended in the chemisorption of the reactants on the surface. An activation barrier to the Langmuir–Hinshelwood reaction is indicated, whose magnitude decreases with the increase of reactant coverage on the surface. This is because of the repulsive interactions that exist between the adsorbates, which raise the energy level of the adsorbed CO and O atoms as coverage increases. CO_2 formation occurs when the CO and O atoms come into contact on the surface. The electronegative O atom acts to reduce the net electron transfer from the metal to the CO molecule, weakening the metal–CO bond and raising the energy of the system. When bond formation between CO and O starts to occur, the energy of the system decreases again. This raising and lowering of the system energy with decreasing CO–O distance constitute the Langmuir–Hinshelwood activation barrier. At low coverages, the activation barrier is measured to be 24.1 kcal/mol, and it decreases to about 12 kcal/mol at high reactant coverage.

The inset of Figure 8.2 suggests a shape for the transition state involved in the catalytic oxidation of CO. Even though the diagram is an obvious oversimplification of the multidimensional reaction trajectory that governs the reaction, the transition state must have a shape something like that suggested. Prior to reaction, the O–C–O "bond" angle is nearly 90°. The product CO_2 molecule has an O–C–O angle of 180°, so the transition state must be characterized by an intermediate angle. This postulated transition state also suggests that asymmetric stretch motion is motion along the reaction coordinate. Measurements of the angular distribution of the product CO_2 show pronounced peaking in the direction normal to the surface, which varies somewhat depending on reactant coverage.[10] This observation, along with translational energy measurements of the desorbed CO_2 product that show excess translational energy in the product molecule,[11] provided tantalizing indications that suggested that detailed studies of

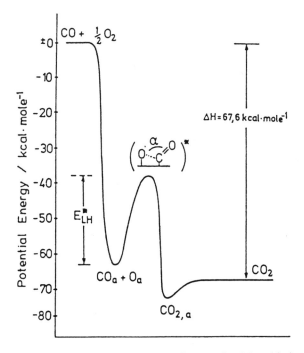

Figure 8.2.   One-dimensional potential energy diagram for CO oxidation on platinum. After ref. 7.

the dynamics of this reaction could provide definitive information about the nature of the transition state in this surface reaction.

Brown and Sibener have used the same experimental approach more recently to study the CO oxidation reaction on the Rh(111) surface. Their work used modulated molecular beam methods to investigate the reaction mechanism and kinetics.[12] In addition, product angular distributions were determined, and TOF measurements of the product translational energy were made.[13] In contrast to the phase and amplitude analysis of the CO oxidation on Pt(111), and as described in the example in Chapter 6, the method used in the study of CO oxidation on Rh(111) collected the entire product waveform. Typical product waveform data for this reaction are shown in Figure 8.3. The rise and decay times are dependent on surface temperature, reactant flux, and reactant beam modulation frequency, and reflect the surface reaction kinetics in terms of a surface transfer function.

The surface transfer function is the frequency space transform of the convolution of several distribution functions that make up the detected waveform indicated in Figure 8.3. Such contributions as the flight time of desorbed product and the residence time of the modulated reactant molecule on the surface are convoluted together in time space to give the measured waveform. In frequency

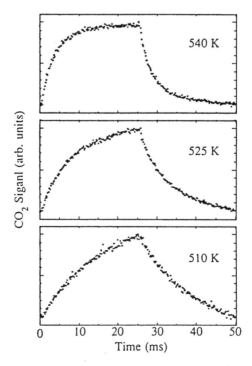

Figure 8.3. Typical product waveform for CO oxidation on Rh(111). After ref. 12.

space, the convolution of distributions is a simple product of their Fourier transforms, so that the surface transform function is made up of the product of the contributing distribution transforms. The reactant molecule residence time is the distribution of interest, as it is the inverse of the rate constant of the surface reaction.

Direct examination of the surface transfer function in frequency space, by plotting the real versus the imaginary part of the transfer function, provides useful information about the surface reaction mechanism. Different reaction mechanisms exhibit distinctive transfer function plots, and mechanistic conclusions can then be reached based on these plots. Transfer function plots for the CO oxidation on Rh(111) clearly indicate that the reaction proceeds by a simple pseudo-first-order process. No evidence for parallel or sequential reaction steps is seen in the transfer function plot for the reaction on Rh(111).[12] This is in contrast to the two-channel reaction mechanism proposed by Segner et al. to explain the CO_2 product angular distribution data for the reaction on Pt(111).[10]

In the Pt(111) case, the angular distribution data was fit by a form $I(\theta) = \alpha \cos^n \theta + (1 - \alpha)\cos \theta$, implying that a fraction $1 - \alpha$ of the desorbing product flux was equilibrated with the surface and desorbed with a $\cos \theta$ angular distribution. The remainder of the desorbing flux was in a highly peaked ($n = 7$) angular distribution, suggesting distinct reaction channels giving rise to the two pieces of the distribution. A similar form can be used to fit the observed product angular distribution data for the Rh(111) surface as well. A reasonable fit is obtained for this data by the equation

$$I(\theta) = 0.65 \cos^{12}\theta + 0.35 \cos \theta \qquad (8.1)$$

This suggests that a two-channel reaction mechanism may be appropriate for Rh(111) as was postulated by Segner et al.[10] for the Pt(111) case.

As mentioned above, however, the shape of the transform function plot argues against a two-channel mechanism on the Rh(111) surface. In addition, direct comparison of the waveforms at $\theta = 0°$ and $\theta = 60°$ detection angles shows no measurable difference. These two detection angles contain primary contributions from the peaked ($n = 12$) and cosine parts of the product angular distribution, respectively. The fact that the waveforms at low surface temperature are indistinguishable strongly suggests identical reaction mechanisms for the product desorbing at these two angles, and that the fit of the measured angular distribution to two terms is likely to be fortuitous.

Further information about the reaction mechanism for CO oxidation on the Rh(111) surface is obtained from direct measurements of the CO_2 product velocity distribution as a function of detection angle.[13] Over the range of surface temperatures from 700 to 1000 K, the measured angular distribution was found to be sharply peaked at the surface normal, and essentially temperature independent. In this higher temperature range, processes relevant to the reaction dynamics can be probed.

Time of arrival waveforms for the product CO_2 were collected for a range of detection angles over the 700–1000 K surface temperature range. In contrast to the observations at low surface temperature, where identical time of arrival waveforms are observed at $\theta = 0°$ and $\theta = 60°$, the velocity of the desorbing CO_2 is seen to decrease with angle from the surface normal in this surface temperature range. The measured distributions appear to be non-Maxwellian, and are actually narrower than would be expected for a Maxwellian distribution with the same average energy. The average translational energy decreases from 8.5 kcal/mol normal to the surface to 4.3 kcal/mol at a desorption angle of 60°. These data are summarized in Figure 8.4. Since the data collection time (6 μsec/ channel) is greater than the surface residence time at these temperatures, the measured differences in product waveforms must certainly reflect the product translational energies.

Even though the measured angular distribution can be well described by the two-channel model proposed by Segner to explain the CO oxidation on Pt(111), no evidence for a bimodality in the measured velocity distributions is observed. Segner's two-channel model would suggest a two-component velocity distribution, particularly at intermediate detection angles where both channels should contribute. This is not what is observed. A single, non-Maxwellian velocity distribution, whose mean energy decreases with angle from the surface normal,

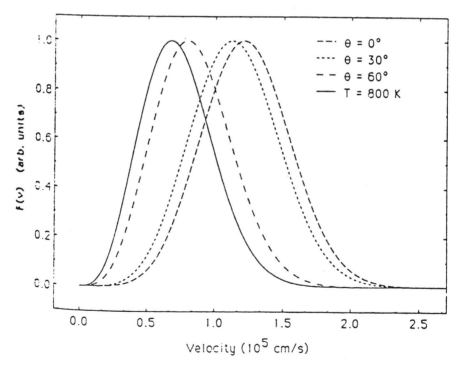

Figure 8.4. Velocity distribution of product CO_2 for various final angles, for CO oxidation on Rh(111). After ref. 13.

is observed. Additionally, if one channel were thermally accommodated, as implied by the Segner model, the large-angle velocity distribution should be Maxwellian at the surface temperature. This also is not observed. In agreement with the low-temperature data, there is no indication that the product flux is generated by two distinct surface processes.

8.2.2 Examination of the Internal Energy of the CO_2 Product of the Catalytic Oxidation of CO on the Platinum Surface

The observation of translationally excited product species suggests that a fraction of the reaction exoergicity is carried away by the product CO_2. The product molecules in this instance certainly do not completely accommodate with the surface prior to desorption. This conclusion is supported by the observation of internally excited CO_2 in the CO oxidation reaction on polycrystalline platinum by Haller and co-workers,[2] Bernasek and Leone,[3] and Brown and Bernasek.[14] Both of these groups used emission methods to detect vibrationally and rotationally excited CO_2 molecules resulting from the oxidation of CO on polycrystalline platinum. Haller and co-workers used an uncollimated free jet expansion of a mixture of CO and O_2 incident on a hot polycrystalline platinum foil, and monitored infrared emission from the desorbed CO_2 resulting from the reaction. The infrared emission was spectrally analyzed by a Fourier transform interferometer and showed substantial rotational and vibrational excitation in the product molecules. Comparison of the resolved emission with calculated spectra, assuming a Boltzmann distribution of levels, suggested a CO_2 rotational temperature of around 1150 K for rotational levels above 25 in the R branch of the 001 to 000 transition. The lower rotational levels in this transition are better fit by a temperature of 400 K. The temperature of the bending mode, again assuming a Boltzmann distribution of energy in the levels, was estimated to be 1750 K. The symmetric stretch temperature was estimated to be somewhat lower, around 750 K. These values for rotational, bending, and symmetric stretching temperatures were then used to synthesize a spectrum for comparison with the measured emission. The asymmetric stretch temperature was varied with the other internal temperature parameters fixed to extract an asymmetric stretch temperature of 2000 K for the best fit comparison.

Bernasek and co-workers used a flow system coupled with various filtering methods to analyze the internal energy of the CO_2 molecules produced in the CO oxidation on a polycrystalline platinum gauze. The apparatus used for these measurements is illustrated in Figure 8.1. A stream of high-purity Ar carrier gas flows over a polycrystalline platinum gauze that can be heated resistively. The reactants are mixed into this flow through a piezoelectric pulsing valve, which provides a modulated stream of reactants to the platinum surface and a modulated emission signal from the product that can be signal averaged. Various filtering methods, including wide and narrow band interference filters, a circular variable filter with a band pass of $\sim$ 30 cm^{-1}, and a cold gas filter cell, were used in these measurements.

Measurement of emission per CO_2 molecule produced indicated that the amount of internal excitation carried away by the product CO_2 molecule increased essentially linearly with increasing surface temperature. Comparison with the emission per molecule observed when pure CO_2 was pulsed over the hot gauze indicated that the reaction product CO_2 molecules were substantially hotter internally than would be expected for thermal accommodation with the surface.

Cold gas filter measurements, where the emission intensity is monitored as a function of room temperature CO_2 pressure in the filter cell, were used to estimate more quantitatively the degree of vibrational excitation in the product CO_2 molecules. Since the room temperature CO_2 in the filter cell will resonantly absorb emission from transitions ending on the (001) or (011) levels, this method provides a means of determining the degree of emission from higher lying levels. Application of this approach to the product emission data over a range of flow conditions and surface temperatures suggests that the average vibrational temperature of the product CO_2 is significantly higher than the surface temperature. For example, at a surface temperature of 1000 K and a flow of 1.5×10^{18} molecules CO/cm^2 over the gauze, the cold gas filter method indicates a vibrational temperature of 1500 ± 100 K for the product CO_2 molecules.

A circular variable interference filter (CVF) with a band pass of about 30 cm^{-1} was used to resolve the emission and provide further details about the internal state of the CO_2 product molecules. Figure 8.5 shows CVF spectra for three different surface temperatures and fixed CO and O_2 flow conditions. The emission per molecule clearly increases with surface temperature, and the spectra are strongly red shifted in each case away from the position of the asymmetric stretch fundamental at 2349 cm^{-1}. This again indicates significant emission from the product CO_2 molecules from high lying vibrationally excited states. Analysis of the CVF spectra allows assignment of asymmetric stretch temperatures for the product molecules, again assuming a Boltzmann distribution of populations. This analysis indicates that the asymmetric stretch vibrational temperatures range from 1100 to over 2200 K for surface temperatures ranging from 800 to 1000 K. Differences between CVF determined asymmetric stretch temperatures and CGF determined average vibrational temperatures, which were especially significant at high surface temperatures and low oxygen flow conditions, indicate that the vibrational modes of the product CO_2 are not in equilibrium with one another.

Vibrational excitation in the product CO_2 was found to increase with increasing temperature or decreasing oxygen flow rate, but to be independent of CO flow conditions. This suggested that the vibrational energy carried away by the product CO_2 species was dependent on the oxygen coverage on the surface during the reaction. A fit of the measured emission per molecule to a simple oxygen coverage model based on surface temperature and oxygen flow resulted in a smooth correlation between product CO_2 emission and relative oxygen coverage over a very wide range of surface temperature and flow conditions. An oxygen coverage dependence of the vibrational energy of product CO_2 was also

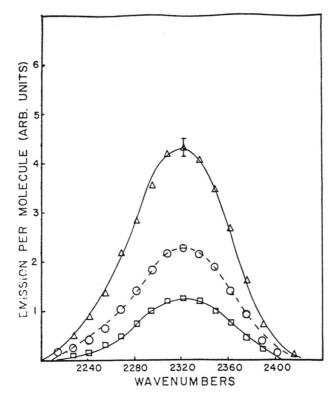

Figure 8.5. Circular variable filter spectra of product CO_2 at three platinum surface temperatures. After ref. 14.

inferred in Haller's group in later time dependent FT–IR emission measurements.[15]

It is interesting to note that the activation energy for this reaction, measured by Campbell et al.[7] and discussed earlier in this chapter, depends on oxygen coverage. Campbell's measurements indicated that the reaction activation energy ranged from 25 kcal/mol at low oxygen coverage to around 12 kcal/mol at coverages near 0.25 monolayer. The observed vibrational energy in the product CO_2 is consistent with a model in which a constant fraction of the activation energy appears as vibrational energy in the product CO_2 molecule with varying oxygen coverage. The observed preference for asymmetric stretch excitation and emission from bend-stretch combination states suggests that an asymmetric stretching motion is motion along the surface reaction coordinate. More extensive coverage-dependent measurements, and higher resolution probing of the product CO_2 internal state distribution is necessary for a more complete understanding of the detailed dynamics of this prototypical surface reaction.

Studies of the internal energy of product CO_2 have recently been extended by both groups. Coulston and Haller[16] have examined the reaction on Pt, Pd, and Rh foils. They observed apparent vibrational temperatures of the product CO_2 that were highest on Pd, less on Pt, and lowest on the Rh surface. On Pd, the vibrational temperature of the product CO_2 was observed to increase with increasing oxygen coverage, in contrast to previous observations on platinum. Relative vibrational populations estimated in this study indicate that the symmetric stretch is more highly excited in the case of the reaction on Pd, suggesting that the activated complex is likely to be more bent on Pd relative to Pt and Rh. Studies by Raines and Bernasek[17] indicate differences in product CO_2 excitation depending on the nature of the oxidant. Both NO and NO_2 appear to result in a more highly excited product CO_2 molecule than when CO is oxidized by oxygen on the polycrystalline platinum surface.

8.3 Comments

The oxidation of CO is probably the most extensively studied catalytic reaction, with a great deal of rather detailed information available about the mechanism and dynamics of the reaction. It has served as a very useful model system for catalytic reactions in general, and has provided most of the dynamic information available about a surface reaction involving more than one molecule. In spite of the many studies that have been carried out, there are still a number of questions about the dynamics of this important reaction that remain unanswered. A number of possible extensions of the studies described in this chapter come to mind that address these questions.

Although the mechanism of the CO oxidation reaction has been studied on well-characterized single crystal platinum metal surfaces, as has the angular distribution and the velocity of the product CO_2, less is known about the detailed internal energy state of the product CO_2 formed on a well-characterized surface. The internal energy measurements described above were carried out on polycrystalline foils or gauzes, and questions remain about the structure and composition of the surface catalyzing the oxidation reaction. The emission methods discussed should be amenable to measurements of the internal energy of the product molecule formed on a well-characterized single crystal. This will be a low signal level experiment, but one that should be possible. The ideal experiment would combine measurement of the angular distribution of the product with translational and internal energy determinations carried out on the same single crystal surface in the same ultrahigh vacuum apparatus. State-resolved velocity measurements using Doppler shift effects may be possible, and would provide a very complete picture of the dynamics of this reaction.

Higher resolution internal energy measurements would also be very useful. While the FT–IR measurements of Haller's group can provide this sort of information, the measurements have not yet been carried out on a well-characterized surface, perhaps because of the difficulty of aligning the interferometer with the desorbing product molecule zone, and the resulting signal limitations. A very promising alternative method for determining the internal energy of the product CO_2 molecule with higher resolution is the use of diode laser absorption spectroscopy.[20] This approach is presently being used in the author's laboratory to examine the state-resolved internal energy state of the product CO_2 molecule. This approach has the resolution to monitor individual rovibrational lines in the CO_2 spectrum, and appears to have the sensitivity necessary to detect product CO_2 formed on a well-characterized single crystal surface in a molecular beam system, as well as CO_2 formed in a flow apparatus similar to that shown in Figure 8.1. The flow system measurements must be coupled with studies in a molecular beam configuration to gain the familiarity with the product CO_2 spectrum that is needed to effectively monitor the low signal level produced when a low surface area platinum single crystal is the active surface. This approach has the advantage that the surface can be prepared and characterized in ultrahigh vacuum, so that the effect of surface structure and composition on the dynamics of this reaction can be appropriately monitored.

The molecular beam scattering approach should also be extended to other single crystal platinum metal surfaces. The differences among the most studied (111) surface and the other low index faces of platinum regarding the dynamics of the oxidation reaction should be determined. Other metals, such as rhodium and palladium, could be examined, and trends in the reaction dynamics determined by the catalyst surface explored. Other oxidants could be used, such as NO, NO_2, N_2O, O_3, and H_2O_2, as suggested by the preliminary studies mentioned above. Differences in the observed internal energy distribution for different oxidant molecules would suggest interesting differences in the mechanism and in the dynamics and transition state of the reaction.

Sufficient detailed dynamic information is now available on the CO oxidation reaction that model trajectory calculations could be undertaken to try to map out the potential energy surface that governs the oxidation reaction, and to get a clearer picture of the transition state associated with the reaction. Studies along this line have been attempted,[16] but further work that takes into account the newer dynamic information that is becoming available should be carried out. This reaction system seems a particularly fruitful one for interaction between experimental dynamicists and theoreticians with expertise in molecular dynamics modeling.

The studies of this chapter also suggest a range of other catalytic oxidation reactions that could and should be examined. The flow system emission and diode laser absorption spectroscopy approach could be used to examine other catalytic oxidation reactions such as NH_3 oxidation, methanol oxidation, and hydrocarbon oxidation. Other reactions catalyzed by platinum metals that form small molecular products that could be readily probed using these methods include the synthesis of HCN, the formation of water or hydrogen peroxide from hydrogen and oxygen, various partial oxidation reactions, and strained ketone or carboxylic acid decomposition reactions that would extrude CO or CO_2. The flow system approach would allow a number of reactions to be screened to determine if sufficient signal would be obtained for emission or absorption measurements of the product internal energy. Promising candidates could then be examined in more detail in the molecular beam reactive scattering configuration, providing the range of detailed information that will enable surface scientists to begin to develop generalized rules for energy disposal and dynamics in surface reactions. In combination with theoretical modeling, a picture of the surface transition state for these important reactions will begin to emerge.

References

1. M. P. D'Evelyn and R. J. Madix, *Surface Sci. Rept.* **3**, 413 (1984).

2. D. A. Mantell, S. B. Ryali, B. L. Halpern, B. L. Haller, and J. B. Fenn, *Chem. Phys. Lett.* **81**, 185 (1981).

3. S. L. Bernasek and S. R. Leone, *Chem. Phys. Lett.* **84**, 401 (1981).

4. J. O. Chu, B. W. Flynn, and R. E. Weston, Jr., *J. Chem. Phys.* **78**, 2990 (1983).

5. C. T. Campbell, G. Ertl, H. Kuipers, and J. Segner, *Surface Sci.* **107**, 207 (1981).

6. C. T. Campbell, B. Ertl, H. Kuipers, and J. Segner, *Surface Sci.* **107**, 220 (1981).

7. C. T. Campbell, G. Ertl, H. Kuipers, and J. Segner, *J. Chem. Phys.* **73**, 5862 (1980).

8. J. L. Gland and V. N. Korchak, *Surface Sci.* **75**, 733 (1978). J. L. Gland, *Surface Sci.* **93**, 487 (1980).

9. J. L. Gland, B. A. Sexton, and G. B. Fisher, *Surface Sci.* **95**, 587 (1980).

10. J. Segner, C. T. Campbell, B. Doyen, and G. Ertl, *Surface Sci.* **138**, 505 (1984).

11. C. A. Becker, J. P. Cowin, L. Wharton, and D. J. Auerbach, *J. Chem. Phys.* **67**, 3394 (1977).

12. L. S. Brown and S. J. Sibener, *J. Chem. Phys.* **89**, 1163 (1988).

13. L. S. Brown and S. J. Sibener, *J. Chem. Phys.* **90**, 2807 (1989).

14. L. S. Brown and S. L. Bernasek, *J. Chem. Phys.* **82**, 2110 (1985).

15. D. A. Mantell, S. B. Ryali, and G. L. Haller, *Chem. Phys. Lett.* **102**, 37 (1983).

16. G. W. Coulston and G. L. Haller, *J. Chem. Phys.* **95**, 6932 (1991).

19. R. Raines and S. L. Bernasek, unpublished work.

20. L. S. Brown, Ph.D. Dissertation, Princeton University, 1986.

Small Molecule
Decomposition Processes

To sort out the complex dynamics of surface reactions involving small mole-cules, identification of stable intermediates and kinetic information about dif-ferent mechanistic pathways becomes essential. In contrast to the previous chap-ters, which emphasized elementary physical processes such as epitaxial growth, energy transfer and diffusion, or very elementary chemical processes such as diatomic bond breaking, or atom recombination, the present chapter addresses much more complex reactive systems. The dynamic understanding of the reac-tions discussed here is not as detailed as the simpler systems considered previ-ously. However, the characterization of reaction intermediates and mechanistic pathways lays essential groundwork for the more detailed dynamic studies that are certain to follow.

9.1 Experimental Methods

Information concerning stable adsorbed reaction intermediates is available from the vibrational spectra of the adsorbed overlayer. Vibrational spectroscopy gen-erally provides a clear chemical identification of adsorbed species, and usually can provide detailed structural information as well. High-resolution electron energy loss spectroscopy (HREELS) is the most widely used surface vibrational spectroscopic method.[1]

In this technique a low energy (2–5 eV), monochromatic electron beam is incident on the single crystal sample. Reflected electrons are energy analyzed, and energy loss peaks in the spectrum can be correlated with vibrational mode excitation in the adsorbed layer. The incident electrons undergo dipole and impact scattering by the molecules or atoms in the adsorbed layer, with loss peaks occurring in the region up to 5000 cm^{-1} away from the elastically scattered peak. The HREELS spectrum provides information essentially equivalent to an infrared spectrum of the adsorbed layer.

Experimentally, this measurement is carried out using an electron spectrometer like the one shown schematically in Figure 9.1. The incident beam is monochromatized by passing through a 127° sector electrostatic analyzer. Typical $\Delta E/E$ for a single pass monochromator of this type is ~0.01. The reflected electrons pass through a similar analyzer, and a spectrum is recorded as electron counts versus analyzer loss energy. This method has very high sensitivity, being able to readily detect coverages on the order of 1% of a monolayer. Resolution is not so good, however, being limited to about 40–50 cm^{-1} FWHM of the elastic peak for a well-tuned single stage monochromator-analyzer instrument. Double stage monochromators and analyzers can be used to improve the resolution to 20–30 cm^{-1} typically, but at considerable loss in sensitivity. Recent advances in instrument design have pushed this to less than 10 cm^{-1}, in the realm of infrared resolution.[2] Typical count rates on the elastic peak for a well-tuned single stage

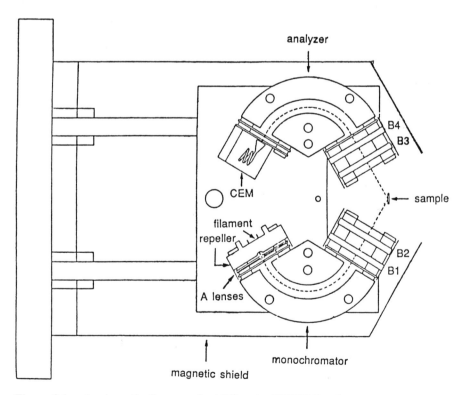

Figure 9.1. A schematic diagram of a 127° sector HREELS optics.

instrument are on the order of 10^6 cps, with loss peak count rates a factor of 10–1000 lower, depending on the nature and order of the substrate. HREELS must be carried out under vacuum to prevent electron–molecule collisions. This is generally not a problem in studies of dynamics on well-characterized single crystal surfaces, but is a concern for high pressure in situ catalytic investigations.

Reflection adsorption infrared spectroscopy (RAIRS) addresses the two major disadvantages of HREELS.[3] The resolution is considerably improved, ranging down to fractions of a cm^{-1} for FT–IR measurements. Also, the photon probe is amenable to high pressure as well as vacuum environments. IR spectroscopy in single reflection mode is considerably less sensitive than HREELS, however, being able to readily detect only strong oscillators down to coverages of a few percent of a monolayer. Experimental improvements in sensitivity are being made continually, suggesting increasing use of RAIRS as a spectroscopic probe of surface reaction dynamics.

In addition to vibrational spectroscopic identification of adsorbed reaction intermediates, it is essential to obtain information about mechanistic pathways and kinetic parameters for those pathways. This information can often be obtained by the use of thermal desorption spectroscopy (TDS), sometimes referred to as temperature-programmed reaction spectroscopy (TPRS).[4] In this method, the overlayer adsorbed on a well-characterized surface is heated so that the surface temperature increases (usually) linearly with time. The partial pressures of desorbing reactants and surface reaction products are monitored as a function of surface temperature. When the desorption occurs into a well-pumped UHV system, the record of partial pressure versus surface temperature shows peaks whose position, intensity, and shape provide information about adsorbed state binding energy, coverage, and desorption or reaction kinetics, respectively. This brief description oversimplifies the situation, of course, and there are numerous pitfalls that must be avoided in the quantitative analysis of thermal desorption data. Descriptions of the technique itself and discussions of data analysis are available in the literature.[4–6] Nevertheless, the method provides a great deal of useful information for understanding surface reaction and desorption processes. Especially when combined with vibrational spectroscopy for intermediate identification, and careful incorporation of isotopic labels, TPRS can be very useful for the elucidation of surface reaction processes.

A final entire class of experimental methods that should be mentioned in the context of small molecule decomposition processes involves those probes using synchrotron radiation to provide structural and identity information for adsorbed layers and reaction intermediates. A number of research monographs are available that describe these techniques, generally and specifically.[7] One technique is described here briefly, as it has direct bearing on the reaction examples provided later in this chapter. That method is near edge X-ray absorption fine structure spectroscopy (NEXAFS)[8] and the fluorescence yield detection modification of this method (FYNES).[9]

NEXAFS monitors the absorption of X-rays by an adsorbed molecular overlayer, in the region near the absorption edge for a particular atomic core level excitation. For example, X-ray absorption would be monitored near the C(1s) or O(1s) core level for molecular overlayers containing carbon or oxygen. Structure in the absorption spectra can be related to the geometric structure of the molecular adsorbate. The absorption process can be followed by experimental measurement of any quantity proportional to the X-ray absorption cross section. Typically, in NEXAFS, total electron emission at energies characteristic of the core level transition being monitored is used to record the NEXAFS spectra. An alternate method of detection is to monitor X-ray photon fluorescence due to the core level excitation, which is again proportional to the X-ray absorption process. This approach, fluorescence yield near edge spectroscopy (FYNES), provides strong signals for C(1s) and O(1s) spectra. It can also be used at high pressures, since it uses only photons as probes.

The classic application of NEXAFS methods to the study of molecular overlayers is given by studies of carbon monoxide, carbonyl, and methoxy species on platinum.[10] The relative intensities of the π and σ resonance absorptions for exciting radiation polarized parallel or perpendicular to the surface normal has been used in these studies to assign a bonding configuration to the various C–O species. Additionally, the energy splitting between σ and π resonances can be related to the C–O bond length in the adsorbed layer.[11]

9.2 Illustrative Examples

A number of examples will be discussed in the remainder of this chapter that illustrate the level of detailed information that is available about the reaction intermediates and pathways for several small molecule decomposition reactions. This information, while not directly dynamic in nature, provides the basis for future studies of the detailed, state-specific dynamics of these reaction systems.

9.2.1 Oxygen Adsorption on Pt(111)

One of the most extensively studied small molecule decomposition reactions is the adsorption and subsequent decomposition of O_2 on the Pt(111) surface. This system, carefully studied by Gland and co-workers[12] provides an excellent example of the interactive use of vibrational spectroscopy and thermal desorption to develop an understanding of a heterogeneous reaction. Thermal desorption spectroscopy of oxygen adsorbed on Pt(111) at low temperatures shows three desorption peaks (Fig. 9.2). These peaks are described as molecular oxygen, desorbing at around 120 K, atomic oxygen, which desorbs following recombination in a broad peak above 700 K, and a peak corresponding to a subsurface "oxide," which begins to decompose around 1250 K.

The surface, when exposed to oxygen at 100 K, exhibits a HREELS spectrum[13] with a strong loss at 870 cm^{-1}. In addition, the work function change of this overlayer was found to be $+0.8$ eV, suggesting electron transfer from the

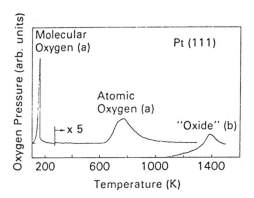

Figure 9.2. (a) Thermal desorption spectrum resulting from exposure of the Pt(111) surface to 5 L of oxygen at 100 K. (b) Thermal decomposition spectrum resulting from exposure of the Pt(111) surface to 100 L of oxygen at 1200 K. After ref. 13.

surface into the adsorbed oxygen. The UPS spectra of this adsorbed layer shows a strong feature about 8 eV below the Fermi energy.[14] Taken together, these observations suggest that the molecular oxygen species is likely to be an adsorbed peroxo species (O_2^{2-}), bound essentially parallel to the Pt(111) surface.

When this layer is warmed slightly, desorption of molecular oxygen occurs around 150 K. Simultaneously, dissociation occurs, and an ordered atomic oxygen overlayer is observed. This conclusion is based upon the disappearance of the 870 cm^{-1} loss, and the growth of a feature at 490 cm^{-1}. The LEED pattern of the surface is a (2×2) structure, even at very low coverages, suggesting formation of oxygen atom islands of (2×2) structure as soon as dissociation occurs. This atomic layer begins to desorb as the temperature is raised above 600 K, in a broad peak indicative of atom recombination.

If the clean surface is exposed to large doses of oxygen while being held at 1000–1200 K, a third type of oxygen overlayer is observed. This species exhibits a strong loss peak at 760 cm^{-1}. This surface is found to decompose with the desorption of oxygen occurring in a mass transfer limited peak around 1250 K. This species appears to be nonreactive below 1100 K, and previous LEED studies[15] suggest that place exchange has occurred and the oxygen forms a subsurface layer. Figure 9.3 summarizes the HREELS spectra for this system, showing the three individual oxygen states, as well as coadsorption overlayers of atomic with molecular, molecular with oxide, and atomic with oxide. These coadsorption spectra, with essentially fixed loss peak positions, suggest that the individual adsorbed states do not affect the local electronic structures of the other adsorbed states.

This series of studies provides an excellent example of the use of spectroscopic methods to determine the various stages involved in a simple heterogeneous reaction. The peroxo species, in particular, is interesting due to the low thermal activation ($\sim$ 40 kJ/mol) needed to convert it to the atomic oxygen

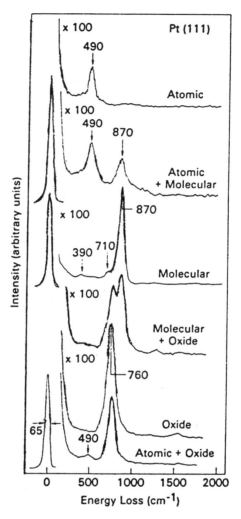

Figure 9.3. A series of energy loss spectra for adsorption and coadsorption of various oxygen species. Coadsorption of molecular oxygen, atomic oxygen, and/or oxide does not significantly perturb the local bonding of any of the oxygen species. After ref. 13.

species. This conversion would be an interesting system for further dynamic investigation.

9.2.2 CO Adsorption and Decomposition on Fe(100)

The adsorption and decomposition of CO on the Fe(100) surface is a second example of a system where reaction pathways and intermediates have been very well characterized. Again, a combination of TDS and vibrational spectroscopy, along with XPS and NEXAFS in this case, has provided a very clear picture of the adsorption and subsequent dissociation of CO on the Fe(100) surface.

Figure 9.4 shows the thermal desorption spectra for a saturation coverage of CO on the Fe(100) surface at 100 K.[16] Three desorption peaks below 440 K and one broad peak at around 800 K are evident in the spectrum. Exposure-dependent

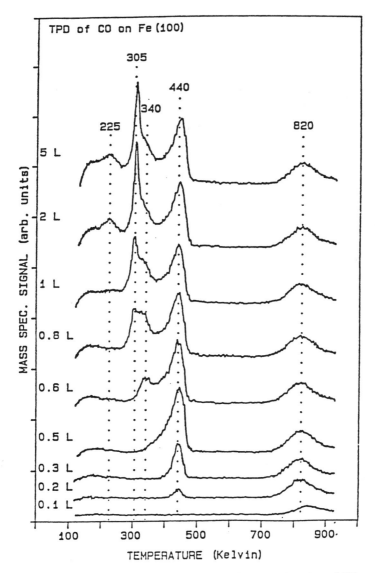

Figure 9.4. TPD spectra of CO on the Fe(100) surface as a function of CO exposure. After ref. 18.

measurements indicate that the peak labeled α_3 saturates first with coverage, followed by α_2 and α_1. The thermal desorption spectrum is shown along with HREELS spectra[17] of the CO overlayer following annealing to the stated temperatures in Figure 9.5. From this figure it is clear that the α_3 state exhibits quite an unusually low C–O stretching frequency of around 1200 cm⁻¹. The α_1 and α_2 states show the more typical ν(CO) frequency near 2100 cm⁻¹.

This information, along with XPS spectra of the O(1s) and C(1s) regions of the CO adlayers,[18] indicates that α_1, α_2, and α_3 are molecular adsorption states, and the state labeled β is due to recombination of adsorbed C and O atoms. It can also be seen from these data that the α_3 state is the precursor to the disso-

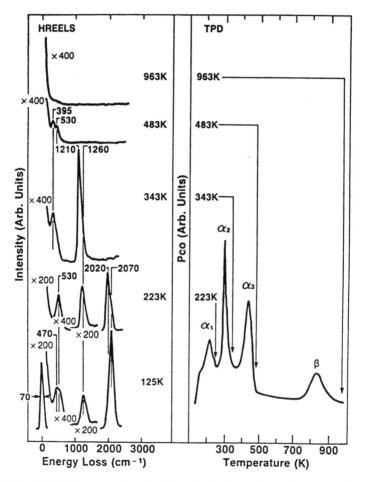

Figure 9.5.   HREELS spectra for the CO adsorbed on a Fe(100) surface after partial CO desorption at the temperature indicated on the reference desorption spectrum. After ref. 18.

ciative β state. Dissociation of CO on this surface does not occur unless the surface is heated through the desorption temperature of the α_3 peak.

The detailed structure of the molecularly adsorbed CO states has also been experimentally determined. NEXAFS measurements[19] on the CO-saturated surface and on the α_3 surface suggested that the α_3 state is bound in the 4-fold hollow site on the Fe(100) surface, tilted at an angle near 55° from the surface normal. This conclusion was based on the nearly identical π to σ resonance ratios that were obtained for normal and glancing incidence geometries for the exciting synchrotron radiation. More recent FYNES measurements provided sufficient signal-to-noise ratio in the C edge spectra to assign normal bonding geometry to the α_1 and α_2 states, and to confirm the proposed geometry of the earlier measurements.[20] In addition, a recent X-ray photoelectron diffraction study[21] derived a tilted geometry for the α_3 state, with the molecular axis inclined 55° with respect to the surface normal, in agreement with the earlier NEXAFS determinations.

These structural determinations also suggested that the C–O bond in the α_3 state had a length typical of nearly a single bond order, consistent with the 1200 cm^{-1} stretching frequency observed in HREELS for this state. Careful, low coverage temperature-dependent HREELS measurements (Fig. 9.6) indicate that the

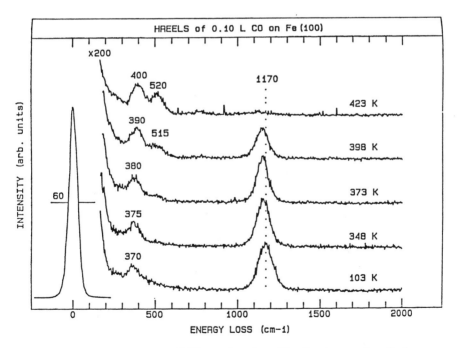

Figure 9.6. HREELS spectra of CO as a function of heating temperature for low coverage. CO exposure for adsorption was 0.1 L. All spectra were taken at 103 K. After ref. 22.

bond lengthens further, as reflected in the slight decrease in $\nu(CO)$, when the α_3 state is heated near to the desorption/dissociation temperature.[22]

The α_3 state forms an ordered c(2 × 2) overlayer on the Fe(100) surface that at saturation amounts to a half monolayer coverage by CO. By careful exposure, or by annealing the CO saturated surface to 400 K, a pure, well-ordered α_3 layer can be produced. When this surface is heated to 500 K, the LEED pattern remains a well-ordered c(2 × 2) structure. However, the HREELS spectrum exhibits peaks at 470 and 530 cm^{-1}, corresponding to M–C and M–O stretching frequencies, respectively.[23] The 1200 cm^{-1} peak of α_3 is completely gone. When the surface is heated above 850 K, the β peak desorbs leaving a clean surface.

All of this information taken together indicates that the dissociation of CO on the Fe(100) surface, by way of the α_3 molecular adsorption state, proceeds as indicated in Figure 9.7. The tilted, lengthened α_3 species, bound with the carbon end down in the 4-fold hollow site of Fe(100), the oxygen bridging two iron atoms along [010] or [001] directions, is the initial stage in this dissociation. As the surface is warmed, the reaction coordinate is along the [010] direction between adjacent iron atoms. The oxygen can move forward over this crest to

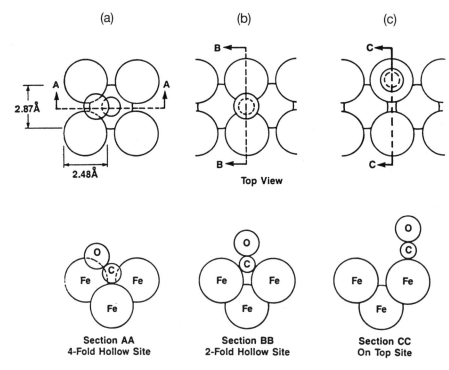

Figure 9.7. Top and cross-sectional views for (a) CO(α_3), (b) CO(α_2), and (c) CO(α_1) on the clean Fe(100) surface. After ref. 18.

an adjacent empty 4-fold hollow site, resulting in dissociation, or it can move back away from the crest, resulting in α_3 desorption.

Again, several spectroscopic methods have been brought together to examine in detail a surface reaction process. The information obtained from these studies provides the basis for carrying out dynamic characterization of this reaction process.

9.2.3 CH₃OH and CH₃SH Decomposition on Fe(100)

As an example of a somewhat more complex small molecule decomposition reaction, consider the interaction of methanol or methanethiol with the Fe(100) surface. The decomposition of these molecules has also been investigated using the combination of TDS and vibrational spectroscopy. In these cases, the ability to monitor several desorbing masses simultaneously in thermal desorption helps to sort out the complex reaction chemistry.

Figure 9.8 summarizes thermal desorption data for methanol adsorption and decomposition on the clean Fe(100) surface,[24] as a function of initial methanol coverage. Masses 2, 16, 18, 28, 30, 31, and 44 have been monitored. The primary

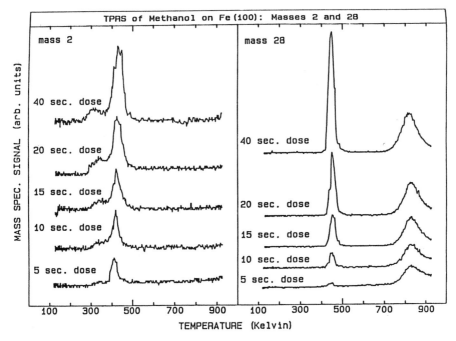

Figure 9.8. TPRS of CH₃OH on Fe(100) as a function of relative CH₃OH exposure for mass 2 (left) and mass 28 (right). After ref. 24.

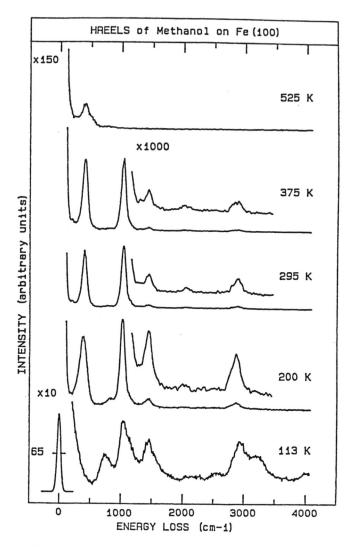

Figure 9.9. HREELS of a saturated overlayer of CH$_3$OH on Fe(100) at 113 K and subsequently heated to higher temperatures. All spectra were taken at 113 K. After ref. 24.

reaction products over the whole coverage range on the initially clean surface are hydrogen and carbon monoxide. In the case of H$_2$, desorption occurs in two peaks, one slightly below room temperature and the second at 440 K. CO desorbs in two peaks as well, the first at the same temperature as the second H$_2$ peak (440 K), and the second near 800 K. Methanol, mass 31, is seen to desorb in a large peak at very low temperature (multilayer molecular desorption) and again in a small peak coincident with H$_2$ and CO desorption at 440 K. Significant amounts of methane ($m/e = 16$), water (18), formaldehyde (30), or carbon dioxide (44) are not observed in the methanol decomposition from the clean Fe(100) surface.

HREELS data for the CH_3OH decomposition on this surface are summarized in Figure 9.9. At low temperatures a spectrum characteristic of a multilayer methanol ice is apparent. Peak positions are in agreement with those observed for condensed methanol, and the OH stretching frequency near 2800 cm^{-1} is evident. When this layer is warmed to desorb multilayer molecular methanol (150 K), an obvious change in the spectrum occurs. The ν(OH) band is lost, and the resulting peaks can be assigned as due to a methoxy species bound upright on the surface, through the oxygen. Deuterium labeling of the methanol confirms these assignments. Essentially no change in the vibrational spectrum is observed as the surface is heated still higher, until the temperature is increased above 440 K, the desorption temperature for the first CO peak and the second H_2 peak. The vibrational spectrum after heating to 523 K shows only low-frequency bands that can be assigned to Fe–O and Fe–C stretching. After heating the surface above 850 K and the desorption of the second CO peak, these bands are absent from the spectrum as well.

Deuterium labeling of the methanol helps to determine the rate-determining

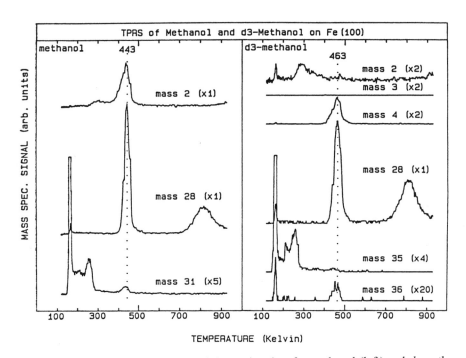

Figure 9.10. Comparison of thermal desorption data for methanol (left) and d_3-methanol (right) on Fe(100). The CO peak evolved from d_3-methanol lies 20 K higher in temperature than that of the methanol overlayer. After ref. 24.

step in the decomposition. This is illustrated in Figure 9.10 where the desorption spectra for hydrogenic molecules, carbon monoxide, and labeled and unlabeled methanol are compared for methanol and d_3-methanol decomposition on the Fe(100) surface. It can be seen that the reaction limited desorption of the first CO peak and the associated hydrogen peak is shifted to about 30 K higher in temperature for the d_3-methanol decomposition. In addition, the hydrogen molecules are separated distinctly into two types in the d_3-methanol decomposition. The low temperature peak is all H_2, suggesting H_2 formation from the alcoholic H made available by initial O–H bond scission and methoxy formation. The higher temperature hydrogen molecule desorption from d_3-methanol is only D_2 (mass 4) suggesting its origin in the methyl hydrogens of methoxy. Similarly, the low temperature desorbed methanol from d_3-methanol is entirely mass 35 (CD_3OH), while that desorbed at 470 K is primarily mass 36 (CD_3OD) resulting from methoxy disproportionation during this rate-limiting decomposition step.

Based on consideration of this data, a schematic picture of the methanol decomposition can be constructed, as illustrated in Figure 9.11a. Initial methanol

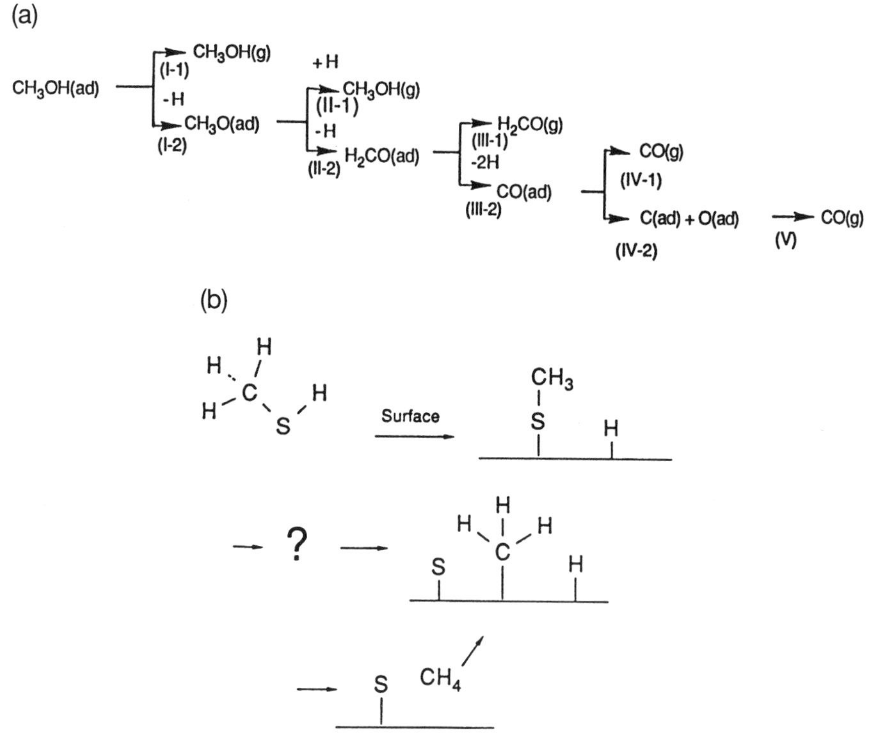

Figure 9.11. (a) A proposed mechanism for reaction of CH_3OH on clean and modified Fe(100). After ref. 29. (b) A proposed mechanism for the reaction of CH_3SH on Fe(100). After ref. 25.

multilayers desorb, followed by O–H bond scission and formation of an upright methoxy intermediate. This intermediate is thermally stable up to 440 K, where decomposition occurs by initial methyl C–H bond cleavage, followed by rapid loss of hydrogen and desorption of CO. Some of the CO further decomposes to form adsorbed atomic C and O, which recombine and desorb near 800 K.

In contrast, the decomposition pathway for methanethiol appears to be controlled by the strength of Fe–S bonding in the reacting overlayer.[25] Exposure at low temperatures again results in a multilayer of intact CH_3SH molecules, as

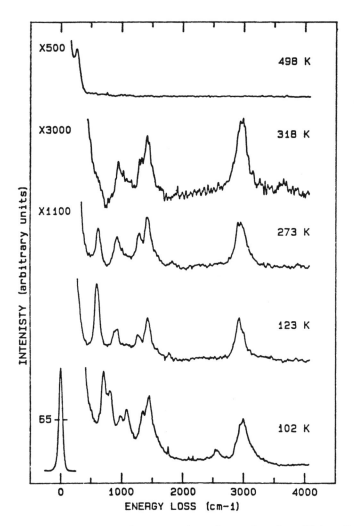

Figure 9.12. HREEL spectra of a saturated overlayer of methanethiol on Fe(100) at 102 K and subsequently heated to higher temperatures. All spectra were recorded at 102 K. After ref. 25.

indicated by the presence of the S–H band at 3600 cm^{-1} and the loss positions in the spectrum shown in Figure 9.12 at 102 K. Heating this overlayer results in molecular desorption at low temperatures, and decrease in the intensity of the S–H band in the HREELS spectra. The species present on the surface at this point can be identified as an upright thiomethoxy intermediate. When this over-layer is heated to 260 K, following desorption of H$_2$ from the surface, but no other reaction products, the HREELS spectrum shows a strong decrease in the ν(C–S) band intensity. The other features, methyl deformation and C–H stretch-ing modes, remain relatively unaffected.

Heating the surface still further initiates the desorption of the major decom-position product, methane. Methane desorbs in a reaction limited peak at 400 K, along with some further H$_2$ product. No sulfur-containing species are seen to desorb from the surface. Accompanying the methane desorption is a decrease in the intensity of the methyl and C–H modes in the HREELS spectra, as seen in Figure 9.12. This data suggest that C–S bond cleavage results in the formation of an adsorbed methyl intermediate, which then recombines with adsorbed hydrogen to form the product methane. XPS spectra of the overlayer as it is heated support this conclusion. In the region where C–S bond scission has occurred and methyl should be present on the surface, the S(2p) XPS spectra are consistent with atomic sulfur bound to the iron surface. The C(1s) spectra, by contrast, remains consistent with a molecularly bound carbon species. This contrasting reaction scheme is summarized in Figure 9.11b for comparison with the methanol decomposition reaction on this surface. Again, these examples provide an idea of the sort of detailed mechanistic chemistry that is beginning to become available for more complex molecular adsorbates on well-character-ized surfaces.

9.3 Comments

The examples presented in this chapter are somewhat different than those dis-cussed in previous chapters. In the first instance, they deal with overall adsorp-tion and decomposition processes rather than an individual elementary step such as diffusion, energy transfer, or diatomic bond breaking and forming. In addition, these examples are concerned with polyatomic molecules in addition to O$_2$ and CO. Finally, the emphasis has been on the identification of reaction intermediates and overall mechanistic processes rather than directly on what has been described as heterogeneous reaction dynamics throughout this book.

Why, then, has this chapter been included in this monograph? Two reasons, primarily. The first is to demonstrate the level of detail that is becoming available concerning the reaction behavior of even rather complex molecular systems on well-characterized surfaces.[26] This is true by no means only of the examples discussed in detail here. Rather, a wide range of molecular reaction processes on a wide range of well-characterized surfaces is being investigated presently.

These studies are laying out the mechanistic groundwork including solid spectroscopic characterization of intermediates and reaction steps, which is essential to the successful pursuit of detailed dynamic investigations.

Which leads to the second primary reason for including the discussion of these particular examples and this chapter in general. In each case, a specific process or intermediate has been observed that deserves further dynamic attention. In the spirit of the Comments sections of the previous chapters, these suggestions for future work are presented. In each case, the suggestions are just that, and do not imply that the particular measurements are definitely possible, or even upon careful consideration, desirable. Hopefully, as in previous chapters, they will serve to stimulate further research.

In the case of oxygen adsorption on the Pt(111) surface, the observation of a highly distorted molecular oxygen species, termed a peroxo (O_2^{2-}) species, is particularly intriguing. It would be of interest to carry out further detailed spectroscopic studies to identify the adsorption geometry, and assign Pt–O and O–O bond lengths with some certainty. In addition, it would be especially interesting to probe the time and temperature dependence of the conversion of the (O_2^{2-}) species to atomic oxygen in real time. This might be possible by monitoring the UPS feature at 8 eV below E_F during rapid laser-induced thermal perturbation of the surface. It would most likely require the UV photon intensity of a synchrotron source to monitor any changes induced by rapid heating. An additional approach might be to use a peroxo coverage modulated by supplying oxygen to the surface with a modulated molecular beam, and again monitor peroxo to atomic conversion using changes in the synchrotron excited UPS spectra.

For CO adsorption on Fe(100), the intriguing observation is, of course, the α_3 molecular adstate. Again, solid spectroscopic information suggests a highly perturbed adsorbed molecule, which is clearly the direct precursor to dissociation of CO on this surface. It would be very interesting, in this instance, to try to probe the internal energy of the CO molecules desorbing from the α_3 state as the surface is thermally modulated across the desorption/decomposition temperature range. Detailed information on vibrational and rotational energy of the desorbing CO molecule, which may be obtainable via laser-induced fluorescence,[27] multiphoton ionization,[28] or other methods (as discussed in Chapters 3 and 7), could help to map out the region of the CO/Fe(100) interaction potential which governs the $\alpha_3 \rightarrow$ C, O decomposition reaction. Again, either thermal modulation by laser heating, or perhaps concentration modulation using a molecular beam, might be employed to carry out such measurements.

The proposed mechanism for methanol decomposition included initial methyl C–H bond cleavage as the rate-limiting step in the decomposition of methanol to CO and H_2. This raises the question about subsequent intermediates in this process. Are CH_2O and CHO species present on the surface in any detectable amount? What is the actual geometry of the methoxy species as methyl C–H bond cleavage occurs? Does the presence of coadsorbates affect the presence and stability of possible H_2CO and HCO intermediates, as is seen for the stability of CH_3O- in the presence of coadsorbed oxygen?[29] These questions would have to be probed using spectroscopic methods with the sensitivity, resolution, and time response necessary to detect what are certainly quite transient, if existent, species. Recently developing second harmonic and sum frequency generation techniques,[30] which appear to have reasonable sensitivity for C–H containing species, may be appropriate in this regard.

The production of CH_4 from the decomposition of CH_3SH appears to be an interesting candidate to probe for excess internal energy in the reaction product. As was seen in the studies considered in the previous chapters, excess internal energy in surface reaction products appears to require weak bonding between the surface and the molecular product. This is certainly the case for the CH_4/ Fe(100) interaction. In addition, the proposed mechanism of CH_4 formation, involving methyl + hydrogen recombination, might be probed by monitoring internal energy in the methane product. There should be some difference in internal excitation between the case in which metal bound methyl recombines with hydrogen, and the one in which hydrogen recombines with methyl in the process of C–S bond cleavage. Again laser-based probes of the desorbing CH_4 would have to be employed to address these questions.

References

1. H. Ibach and D. L. Mills, *Electron Energy Loss Spectroscopy and Surface Vibrations,* Academic Press, New York, 1982.

    G. Ertl and J. Kuppers, *Low Energy Electrons and Surface Chemistry,* VCH, Weinheim, 1985.

2. H. Ibach, M. Balden, D. Bruchmann, and S. Lehwald, *Surface Sci.* **269/270,** 94 (1992).

3. Y. J. Chabal, *Surface Sci. Rept.* **8,** 211 (1988).

4. R. J. Madix, *CRC Crit. Rev. Solid State Materials Sci.* **7,** 143 (1978).

5. A. M. deJong and J. W. Niemantverdriet, *Surface Sci.* **233,** 355 (1990).

6. D. A. King, *Surface Sci.* **47,** 384 (1975).

7. G. Margaritondo, *Introduction to Synchrotron Radiation,* Oxford University Press, New York, 1988.

8. J. Stöhr and R. Jaeger, *Phys. Rev. B* **26,** 4111 (1982).

9. D. A. Fischer, U. Dobler, D. Arvanitis, L. Wenzel, K. Baberschke, and J. Stöhr, *Surface Sci.* **177,** 144 (1986). F. Zaera, D. A. Fischer, S. Shen, and J. L. Gland, *Surface Sci.* **194,** 205 (1988).

10. F. Sette, J. Stöhr, and A. P. Hitchcock, *Chem. Phys. Lett.* **110,** 517 (1984).

    F. Sette, J. Stöhr, and A. P. Hitchcock, *J. Chem. Phys.* **81,** 4906 (1984).

11. J. Stöhr, J. L. Gland, W. Eberhardt, D. Outka, R. J. Madix, F. Sette, R. J. Koestner, and V. Dobler, *Phys. Rev. Lett.* **51,** 2414 (1983).

12. J. L. Gland, *Surface Sci.* **93,** 487 (1980).

13. J. L. Gland, B. A. Sexton, and G. B. Fisher, *Surface Sci.* **95,** 587 (1980).

14. G. B. Fisher, B. A. Sexton, and J. L. Gland, *J. Vac. Sci. Technol.* **17,** 144 (1980).

15. G. Kneringer and F. P. Netzer, *Surface Sci.* **49,** 125 (1975).

16. D.-W. Moon, D. J. Dwyer, and S. L. Bernasek, *Surface Sci.* **163,** 215 (1985).

17. D. W. Moon, D. J. Dwyer, J. L. Gland, and S. L. Bernasek, *J. Am. Chem. Soc.* **107,** 4363 (1985).

18. D. W. Moon, J.-P. Lu, D. J. Dwyer, J. L. Gland, and S. L. Bernasek, *Surface Sci.* **184,** 90 (1987).

19. D. W. Moon, S. Cameron, F. Zaera, W. Eberhardt, R. Carr, S. L. Bernasek, J. L. Gland, and D. J. Dwyer, *Surface Sci. Lett.* **280,** L123 (1987).

20. D. J. Dwyer, B. Rauersenberger, J.-P. Lu, S. L. Bernasek, D. A. Fischer, S. D. Cameron, D. H. Parker, and J. L. Gland, *Surface Sci.* **224,** 375 (1989).

21. R. S. Saiki, G. S. Herman, U. Yamada, J. Osterwalder, and C. S. Fadley, *Phys. Rev. Lett.* **63,** 283 (1989).

22. J.-P. Lu, M. R. Albert, and S. L. Bernasek, *Surface Sci.* **217,** 55 (1989).

23. J.-P. Lu, M. R. Albert, C. C. Chang, and S. L. Bernasek, *Surface Sci.* **227,** 317 (1990).

24. M. R. Albert, J.-P. Lu, S. L. Bernasek, and D. J. Dwyer, *Surface Sci.* **221,** 197 (1989).

25. M. R. Albert, J.-P. Lu, S. L. Bernasek, S. D. Cameron, and J. L. Gland, *Surface Sci.* **206,** 348 (1988).

26. S. L. Bernasek, *Annu. Rev. Phys. Chem.* **44,** 265 (1993).

27. R. N. Zare and P. J. Dagdigian, *Science* **185,** 739 (1985).

28. P. M. Johnson, *Acc. Chem. Res.* **13,** 1 (1980)

29. J.-P. Lu, M. R. Albert, S. L. Bernasek, and D. J. Dwyer, *Surface Sci.* **239,** 49 (1990). J. P. Lu, M. R. Albert, S. L. Bernasek, and D. J. Dwyer, *Surface Sci.* **218,** 1 (1989).

30. Y. R. Shen, *Annu. Rev. Phys. Chem.* **40,** 327 (1989).

Index